ENGINEERING PHILADELPHIA

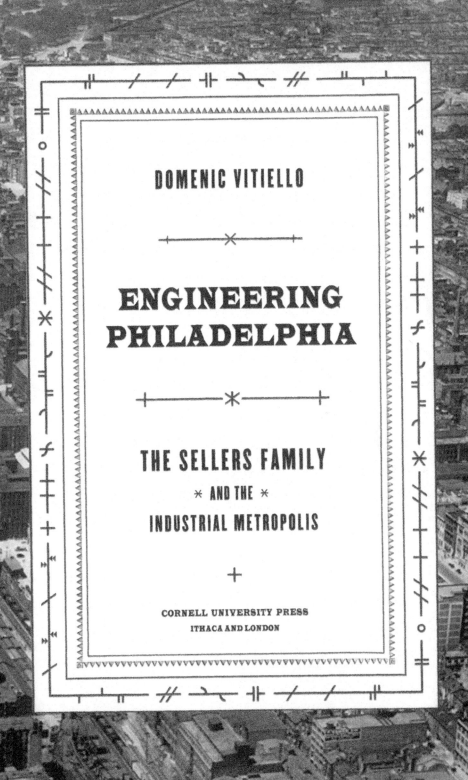

DOMENIC VITIELLO

ENGINEERING PHILADELPHIA

THE SELLERS FAMILY
* AND THE *
INDUSTRIAL METROPOLIS

CORNELL UNIVERSITY PRESS
ITHACA AND LONDON

First published 2013 by Cornell University Press

Printed in the United States of America

Library of Congress Cataloging-in-Publication Data
Vitiello, Domenic, author.
    Engineering Philadelphia : the Sellers family and the industrial metropolis / Domenic Vitiello.
        p. cm.
    Includes bibliographical references and index.
    ISBN 978-0-8014-5011-2 (cloth : alk. paper)
    1. Sellers family. 2. Philadelphia (Pa.)—History—19th century.
3. Philadelphia (Pa.)—Economic conditions—19th century.
4. Manufactures—Pennsylvania—Philadelphia—History—19th
century. 5. Industrialization—Pennsylvania—Philadelphia—
History—19th century. 6. Urbanization—Pennsylvania—
Philadelphia—History—19th century. 7. Deindustrialization—
Pennsylvania—Philadelphia—History—20th century. I. Title.
    F158.44.V64   2013
    974.8'11—dc23   2013012347

Cornell University Press strives to use environmentally responsible suppliers and materials to the fullest extent possible in the publishing of its books. Such materials include vegetable-based, low-VOC inks and acid-free papers that are recycled, totally chlorine-free, or partly composed of nonwood fibers. For further information, visit our website at www.cornellpress.cornell.edu.

Cloth printing   10 9 8 7 6 5 4 3 2 1

CONTENTS

List of Illustrations   *vii*

Preface   *ix*

Introduction   1

1. Manufacturing Metropolitan Development   11

2. Migration Strategies and Industrial Frontiers   46

3. Rationalizing the Factory and City   76

4. Progressive Economic Development   107

5. Empires of Steel   136

6. Building the Scientific City   156

7. Roots of Decline   192

Notes   *217*

Index   *259*

ILLUSTRATIONS

1. Portrait of Escol Sellers, c. 1897.   3

2. Charles W. Peale sketch portrait of Nathan Sellers, c. 1810.   13

3. Detail of Scull and Heap's map of the Philadelphia region, c. 1777.   15

4. Sellers & Pennock advertisement showing the Hydraulion, c. 1818.   36

5. Advertisement for Coleman Sellers & Sons, c. 1839.   39

6. Map of the Bellefontaine & Indiana Railroad, 1853.   73

7. William Sellers & Co. machine shop, c. 1895.   80

8. Railroad lines encircling central Philadelphia, 1862.   84

9. Fire insurance atlas of the Bush Hill district in 1875.   86

10. Lithograph of William Sellers & Co. factory, c. 1868.   87

11. A still frame from a motion picture Coleman Sellers took in 1859.   94

12. A Sellers & Co. steam hammer, c. 1890s.   97

13a. Portrait of William Sellers, c. 1880.   108

13b. Portrait of Coleman Sellers, c. 1895.   109

14. University of Pennsylvania power plant, c. 1900.   132

15. A drawing of the Midvale Steel works in 1879.   143

16. A forty-ton crane at the navy yard at Portsmouth,
    Virginia, c. 1895.   150

17. Breech-loading rifles at Midvale Steel, c. 1901.   152

18. Edge Moor's Susquehanna River Bridge, c. 1885.   161

19. Edge Moor's market house at Demerara, Guiana, 1882.   164

20. Plan of Fairmount Park, 1870.   170

21. Atlas of the area around the Pennsylvania Railroad
    abattoir, 1901.   173

22. Drawing of Ridley Park, 1872.   179

23. Detail of a map of Ridley Park, 1889.   180

24. Atlas showing the Sellerses' Clifton estate, 1893.   184

25. Cartoon of William Sellers, c. 1900.   198

26. Aerial photograph of Bush Hill in 1925.   201

PREFACE

+———————————————————————————————————+

This book results from a fascination with what made America's industrial cities so prosperous in the past, and with why they declined so precipitously in the twentieth century. Specifically, how could Philadelphia capitalists carve out such a prominent place in the economic geography of the nineteenth century and then fail to sustain this position? Urban historians have explored the effects (more than the causes) of economic decline, tracing how the flight of people and capital set cities like Baltimore, Detroit, and Philadelphia on a downward spiral, especially after World War II. We have done less to explain how people organized the economic development of these places in the nineteenth century and then what went wrong in the early twentieth.

Most historians study economic change through the story of a single company, institution, or sector. By contrast, this book takes the form of a biography of a family of millers, mechanics, manufacturers, engineers, and a corporate titan or two, the Sellers family, descendants of an English farmer and weaver of wire who landed in Philadelphia in 1682. Unlike more traditional biographers, I am concerned less with the family's personal lives and more with what its members illuminate about the city and region around them. Following mostly the men in the family through various realms of public and private life offers a way to trace the overlapping networks of people, businesses, and institutions through which different interests aligned to influence the city's development across distinct chapters of its history.

This is a metropolitan study, a frame of analysis that captures two interrelated scales and spheres of activity: the individual urban region and the connections and interaction between regions. Following the Sellers family makes this mainly a story about Philadelphia. However, they also traveled, did business, built infrastructure, and in some cases resided in other cities of the Northeast, Midwest, South, Europe, Latin America, and Australia. Some of their inventions, plans, and investments helped establish formative patterns of urban and economic development. Hence, this is also a broader story about the ways in which national and international networks—of people, businesses, institutions, information, and infrastructure—emerged and evolved to create industrial cities and define their place in world.

In the broad scope of their public and private pursuits, the Sellerses illuminate the planning, growth, and decline of industrial cities. The engineers in the family reflect a common pattern among nineteenth-century manufacturers and elite engineers, a sustained effort to influence the course of urban economic, social, and physical development around them. Like other capitalists and designers of the material world, they were deeply engaged in urban planning before it became a profession in the twentieth century. They played important parts in the eighteenth- and nineteenth-century movements and institutions that established agendas for urban development and reform and thereby became the intellectual foundations of early city and regional planning. Through varied interventions of economic development—that is, by planning—they sought to impact "economic development" in a broader sense, through the growth of industrial capitalism.

Family members' work in firms and institutions covered the diverse dimensions of metropolitan economic development (in both planners' and economists' terms). These included involvement in developing technologies, businesses, entire sectors of industry, factories, infrastructure, and real estate, as well as leadership of institutions that influenced education, labor, public policy, and the relationships between different sectors. Economic development meant different things across generations, and the roles of the Sellerses and allied capitalists shifted consequently. What remained constant for some two centuries, though, was their concerted effort to shape these various dimensions of public and private life, usually all at the same time.

As with most families, the Sellerses' record has strengths and silences, shining greater light on particular people, places, firms, organizations, and issues in different eras. The continuous analytical thread that we can follow concerns how the engineers in the family sought to determine the course of economic planning and growth in Philadelphia and other cities across more than two centuries. I have chosen to tell the family's story principally around four individuals—Nathan, Escol, Coleman, and William Sellers—as their records and life stories intersect the greatest number and scope of business and institutional activities. They were also, not coincidentally, the most important engineers in the family, and those most deeply engaged in civic institutions and in major projects of technology and infrastructure.

The evidence for much of the book exists thanks partly to Samuel Clemens and especially his friend, Hartford newspaperman Charles Dudley Warner, who first incited Escol Sellers and his nephew Horace Wells Sellers to collect, write, and preserve their family's history. The introduction and chapters 1 and 2 are based on Escol's memoirs and especially the trove of letters and notebooks, including those of his grandfather Nathan, that Horace assembled beginning in the late nineteenth century.[1] Escol's youngest brother, Coleman, and their second cousin William are the central characters of the remainder of the book. Their stories are based less on personal records, most of which apparently do not survive, and more on the ledgers of companies, engineering literature, and minutes of institutions. Maps, deeds, and other real estate records help illustrate the metropolitan geography shaped by each generation.

The chapters of the book survey the ways that mechanics, surveyors, and engineers sought to and did influence economic and urban development at various scales and across different eras of history. In the late seventeenth and eighteenth century, Nathan and his father and grandfather's generations illuminate how Philadelphians tied urbanization to industrialization, organizing to shape entire sectors, economic and social policy, and the infrastructure to support a diverse and dispersed metropolitan economy. In the early nineteenth century, the Sellerses played important roles in the transition to industrial ways of work, production, and city building. Chapter 1 examines how these manufacturers and surveyors established patterns of business, institutional, and urban development that made Philadelphia a metropolitan center with a prominent—and distinctly industrial—place in the colonial Atlantic and early national economy. Chapter 2 follows Escol's generation as they sought to shape the expansion of the United States, as urbanization and industrialization—and their relative absence and failures in certain regions, especially the South—defined a new economic geography in the decades before the Civil War.

William and Coleman's careers in the second half of the nineteenth century show how engineers and their allies developed formative technologies, theories, and methods of planning and building industrial cities. Chapter 3 examines their machine-tool works and its impacts on the rationalization of factories and urban growth in the mid-nineteenth century. Chapter 4 traces the complex of institutions through which the Sellerses and their allies influenced regional economic development and pre-Progressive urbanism during and after the Civil War. Chapter 5 explores William's postwar ownership of Midvale Steel and how it influenced new corporate patterns of industry and economic development, including the rise of the military-industrial complex and national capitalism. Chapter 6 charts the Sellerses' diverse and often lasting impacts on the metropolitan built environment in the late nineteenth century.

Some of these broad patterns of industrialization and urbanization that the Sellers family helped set in motion contributed to Philadelphia's decline in the twentieth century. Despite the considerable capital and influence they acquired in the eighteenth and nineteenth centuries, after 1900 the Sellerses and other local industrialists largely lost the power to drive their region's economic development. The final chapter relates how they sold their companies and how they and other capitalists reconfigured the city and its institutions and economy in ways that reflected and helped define the character and course of its industrial decline. The interaction of macroeconomic forces with local decisions and actions highlights some familiar and some less-explored causes and reactions to metropolitan and industrial restructuring.

Ultimately, this is a story about the diverse ways in which people seek, succeed, and fail to influence the prosperity and poverty of cities and their position in larger national and world economies. It is an urban and planning history, concerned with the social and spatial dimensions of economic development at least as much as the business strategies and technologies that most historians of industrialization have examined. The Sellerses elucidate the biography of industrial Philadelphia and many of the places to which it was linked, beyond the fortunes of an individual firm, district, or sector. In this sense, the metropolis itself is the central character of this book.

Many people generously supported this project. As generations of urban historians have recognized, Michael Katz is a model mentor. I consider myself extremely fortunate to continue to work with him as we study cities together at the University of Pennsylvania. Walter Licht's masterly teaching and advice enabled me to become fluent in industrial history while remaining focused on big questions of social import. John K. Brown's scholarship helped inspire the research for this book, and I am deeply appreciative of his close reading and detailed feedback. George E. Thomas first introduced me to the work of William Sellers and the links between mechanical engineering, architecture, and urbanization. Andrew Dawson, Howell Harris, Thomas Heinrich, Thomas Hughes, David Meyer, Phil Scranton, and Ross Thomson offered advice that illuminated the Sellerses' place in the broader context of American industrial history. Lily Milroy did likewise for their place in the history of environmental planning. Ellen Cronin, Burton Sellers, Nicholas Sellers, and Peter and Lucy Bell Sellers helped fill out various chapters of the family's history.

In the archives, Robert Cox, Roy Goodman, Valerie-Anne Lutz, and their colleagues made it a pleasure to get to know the Sellerses through their letters and materials at the American Philosophical Society. The staff at the Soda House and Library at Hagley graciously accommodated much of my research on the Sellers firms; and Marjorie McNinch also directed me to sites associated with the Sellerses and their relations in Delaware. Virginia Ward helped me explore the Franklin Institute's little-used but tremendously useful collections. Tom Smith shared the scrapbooks of the Sellers Family Library as well as his own work on the family. Theresa Stuhlman aided my research at the Fairmount Park Commission archives. Richard Boardman at the Free Library of Philadelphia Maps Department has graciously assisted me in countless hours of poring over fire insurance atlases for this book and much related research.

Thanks also to the staffs of the City of Philadelphia Archives; the Free Library Periodicals Department; the Friends Historical Library at Swarthmore College; the Historical Society of Delaware; the Historical Society of Pennsylvania; the Library Company of Philadelphia; the National Archives in Philadelphia and Washington, DC; and the Pennsylvania State Archives. Thanks as well to Mark Lloyd and his staff at the University of Pennsylvania Archives, and John Pollack at the University of Pennsylvania's Van Pelt Library Special Collections.

At Cornell University Press, Michael McGandy has been a fabulous editor, managing a tremendously helpful peer review and offering incisive suggestions of his own. His colleagues at the press, including Susan Barnett, Sarah Grossman, and Susan Specter, as well as the copy editor, Glenn Novak, and the indexer, Dave Prout, provided outstanding support and attention to detail in producing the book. I am also deeply appreciative of the generous critiques and advice of the anonymous readers for the press.

Two people deserve the lion's share of the credit for the images in this book: Ashley Hahn helped select and assemble photos and maps that illustrate many parts of the Sellerses' and Philadelphia's story. My dean at the University of Pennsylvania, Marilyn Jordan Taylor, provided a generous grant from the Grosser Fund to dramatically expand the visual dimension of the book.

I also thank colleagues at the University of Pennsylvania who have influenced my thinking about economic development and urban and planning history, and who have supported me in myriad other ways, including Eugenie Birch, Tom Daniels, Amy Hillier, David Hsu, John Landis, and Laura Wolf-Powers in the Department of City and Regional Planning, and Michael Katz, Andrew Lamas, Eric Schneider, Elaine Simon, Mark Stern, and Tom Sugrue in Urban Studies.

Finally, I thank my wife, Soumya Iyer, for critiquing an early draft of this book, and for everything else she does to support my work; and our young son, Luca Iyer Vitiello, for napping long enough to let me finish it.

ENGINEERING
PHILADELPHIA

INTRODUCTION

———————————————✕———————————————

"There's millions in it!" exclaimed Colonel Eschol Sellers as he conjured up schemes to sell patent medicines, build railroads, corner the market on hogs, and turn remote lands in Tennessee into cities buzzing with enterprise. The protagonist of Mark Twain and Charles Dudley Warner's novel, *The Gilded Age*, the Colonel was a sympathetic but delusional character. His tongue, they wrote, "was a magician's wand that turned dried apples into figs and water into wine as easily as it could turn a hovel into a palace and present poverty into imminent future riches."[1] His plans invariably rested on the corrupt and empty promise of appropriations from Congress. For Twain, Warner, and generations of readers, Colonel Sellers personified the culture of blind speculation in the economic booms (and busts) of nineteenth-century America.

For the real Escol Sellers, the authors' use of his name was infuriating, as was their interpretation of industrial capitalism.[2] He was indeed the inspiration for the name, as Warner knew of him through a mutual associate.[3] Escol was deeply offended in part since he was a very different man, a well-respected figure in American engineering with numerous patents to his name. Unlike Twain and Warner's Sellers character, he and his family actually did build industries, infrastructure, and cities.

Unfortunately, however, portions of Escol's career resembled that of the fictional Colonel, which must have fueled his embarrassment. After moving to Cincinnati in the 1840s, he consumed himself with promoting light railroad systems to reach California and to cross the Isthmus of

Panama, and subsequently sought to make coal gas and paper in Illinois. None of these ventures enjoyed much success. More damningly, in the 1840s and early '50s, Escol and his brother Charles developed iron forges in Tennessee and railroads in Georgia, backed by the infamous southern industrial promoter General Duff Green and using slave labor. An early lobbyist in Washington, Green was one of the men responsible for the Crédit Mobilier corporate scandal, the Enron of its day. Indeed, the collapse of Crédit Mobilier in 1872 did more than a little to influence Twain and Warner to write their biting satire of American capitalism the following year.

On several occasions Escol vowed to sue the book's publisher, but instead he ended up writing his own version of the story. Already an old man, he retired to his study in a house he bought on a hill overlooking Chattanooga, Tennessee. He spent the remaining two decades of his life penning his reminiscences. The premier engineering journal of the era, *American Machinist*, published lengthy excerpts between 1884 and 1895. Historians of engineering consider it one of the most important first-person accounts of early American industrialization.[4]

Escol's memoirs revealed what was at stake in his disagreement with Twain and Warner. Beyond his own bruised ego, he took issue with their portrayal of how industrial America was built. Eschewing their cynicism, he celebrated the roles of engineers in transforming nineteenth-century society and material life. As a response to *The Gilded Age*, Escol's account may be read as a detailed corrective on the process, geography, and moral character of industrialization.

His story started in Philadelphia, not the pioneer West or Washington, and described how the city became a place that mattered in the world's early industrial economy. At its center were the men who led the firms and institutions of the city and the principal manufacturing hubs of Britain. He recalled his boyhood in the early nineteenth century, playing and learning in his grandfather Charles Willson Peale's museum, which popularized natural history among the city's small middle class. He filled many pages detailing the life around his father's workshop, where a tight-knit community of producers and scientists gathered to talk, experiment, and collaborate in inventing a new world of mechanized factories and infrastructure. Their business networks and institutional projects did much to shape economic development (meaning both planning and broader economic growth) within as well as between urban regions.

Escol cast industrialization as a sincere pursuit of public import. He wrote extensively about his first trip to Britain in the early 1830s to gain new knowledge and transfer technologies, but tellingly left out most of his

Figure 1. Portrait of Escol Sellers late in life, by George Giguere. In Eugene S. Ferguson, ed., *Early Engineering Reminiscences (1815–1840) of George Escol Sellers* (Washington, DC: Smithsonian Institution, 1965), after *Paper Trade Journal* (October 16, 1897), vol. 26, 104.

later forays in the American South and West. Of his grandfather Nathan, he barely mentioned his key role in the history of energy but elaborated on the ways he manipulated his neighbor to give up alcohol and thereby escape poverty. A mainly Quaker family, the Sellerses generally sought to engineer social and material life to build what they viewed as a moral economy, with the important exceptions of their southern projects, which family members struggled to rationalize.

The Sellerses' story engages the various dimensions of economic development planning and practice, a large but relatively unexplored area of urban and planning history. It reveals an evolving set of overlapping social, technological, business, and institutional networks, which, understood together, afford an appropriately complex and dynamic portrait of the ways in which people planned and implemented urban economic development—the set of planning and allied interventions as well as the broader growth and evolution of the metropolitan economy. The Sellerses added value to products, diversified regional economies, planned and built dense and dispersed networks of businesses and infrastructure, connected different sectors of the economy, and influenced public policy. In short, they organized the growth of the industrial metropolis. Their experiences also highlight the power struggles, social tensions, and uneven geography of industrialization and urbanization. Despite and often because of this tension and competition, engineers and their allies produced lasting and sometimes seminal technologies, institutions, and paradigms of economic development and urbanization.

Giving Economic Development a (Pre-)History

Although the Sellerses identified principally as mechanics (later engineers) and manufacturers, they played all sides of economic development. Planners define economic development as a broad category of interventions encompassing the support of firms and sectors, labor force training and deployment, real estate and infrastructure investments, research and innovation, diffusion of technologies, and promotional activities. In the second quarter of the twentieth century, city governments and academics unified these tasks under the title of "economic development," theorizing their need to correct market failures and reap broader social benefits from the actions of firms. Practitioners and historians of economic development planning have rarely recognized industrialists' earlier interventions as a larger practice of economic development.[5] Yet taken together, they constitute just that, and they help expand the scope of planning history before the twentieth century beyond questions of physical design.[6]

Locating economic development in the "prehistory" of the profession, before the twentieth century, requires attention to people who did not call themselves planners but nonetheless mediated urbanization and industrialization. Many historians and social scientists have stressed the utility of mediators (of various sorts) for understanding how social and economic change unfolds.[7] Tracing the ways people employed firms, institutions, and infrastructure to organize regional economic development reveals the social processes that mattered at least as much to shaping cities and industrial capitalism as money, technology, and firms.

Making the most of what mediators can reveal about economic development requires modes of inquiry distinct from historians' traditional biographies of a single individual or business. Similarly, describing how the different dimensions of economic development (meaning both planning and capitalism) evolved does not necessarily make sense of the dynamic links between, for example, the education of workers and the promotion of a region's position in the world economy. Instead, historians can capture the broader workings of economic development at the metropolitan or larger scales and the visions and paradigms it advances in particular eras by following especially active and well-connected people through firms, institutions, and projects, and by focusing on the relationships between them.

American industrialization and the integration of dispersed regions' economies largely resulted from the deliberate, systematic work of engineers like the Sellerses. Beginning in the colonial era, they helped build a class of mechanics and manufacturers with the specialized knowledge, money, and power to influence the growth of entire sectors and cities. In the process, they helped define other classes of high- and low-skilled workers, remaking the relationships between capital and labor, and between home and work. The tensions inherent in these transformations suggest some of what was at stake for different groups of people with divergent visions of American capitalism.

Economic development meant different things at different times, scales, and settings. The regional and national landscapes of social and economic interests, the terms of debate, and paradigms of economic development shifted across the sweep of urban and economic history. In the eighteenth century, craftspeople like the Sellerses enabled a distinctly industrial mode of colonization that supported not only economic growth and diversification but also the related projects of nation building and social reform during and after the Revolution. The transition to wage labor and public schooling, mechanized factories and infrastructure, and mineral energy in the early nineteenth-century Northeast created a larger and more diverse industrial economy and society. In the antebellum era, mechanics from the East joined fellow capitalists in the race to industrialize and integrate the cities and regions of a fast-expanding and increasingly fractious nation.

As industrialists confronted booming cities and sought international markets in the second half of the nineteenth century, their institutions and methods of urban and economic development became increasingly technical and scientific. At the regional level, mid-nineteenth-century Philadelphia's technologies and politics of industrial expansion tied imperial economic ambitions to the rationalization of industry and of urban governance and growth. During and after the Civil War, William and Coleman Sellers and their associates created institutions that would give rise to an early Progressive urban reform agenda with corresponding professions, academic disciplines, and real estate interventions that came to define modern American urbanism and urbanization.

In the same period, they developed national corporations and institutions that served the United States' global commercial and military expansion in the late nineteenth century. Their firms and technologies played early and large roles in shaping the national capitalism of what historians term the second industrial revolution. These changes would ultimately redefine their city's place in the world economy, in ways mostly beyond their control.

The Sellerses and their contemporaries in the twentieth century responded to and mediated the Philadelphia region's decline in ways that highlight an important and little-studied chapter of urban history. Scholars have thoroughly examined planners' responses to deindustrialization after World War II and the ways in which midcentury redevelopment failed to regrow urban industries and even exacerbated social and economic problems. They have rarely traced the early twentieth-century roots of decline or the causal roles that planners and local capitalists sometimes played, often inadvertently, in dismantling firms, institutions, and networks.[8]

A family of capitalists can be an ideal unit of analysis to explore the broad dynamics of economic development. A family like the Sellerses affords an intimate view of the networks of people, companies, and institutions that built industrial Philadelphia and other places across generations. Following them through their diverse activities and interactions yields a fuller understanding of the larger process of metropolitan economic development than the history of a single firm or institution affords.

Overlapping Networks

The Sellerses were embedded in and purposefully cultivated overlapping networks that influenced economic development. In their businesses, they formed networks of production at the regional and transatlantic scales, and they created the technological systems that organized factory production as well as metropolitan infrastructure networks. Their networks of research and innovation in technology, management, and construction extended to a variety of scientific, educational, and public sector institutions. Through early trade and professional associations, elite social reform movements, and networks of family and religious affiliation, they used their personal connections, more or less instrumentally and successfully, to further particular visions and projects. The extended family's travels to live, work, or do business in the United States and other parts of the world also suggest some of the ways in which their social and economic networks shaped and were shaped by networks of communication, transportation, and trade between cities.

Much historical and social science literature has emphasized the importance of tracing networks to understand society and economy in something resembling their great complexity.[9] American planning historians have examined networks of urban and economic development by researching the growth of industrial districts. They borrow from political scientists the analytical framework of the "growth machine," tracing the alliances between manufacturers, real estate developers, politicians, and others invested in urban and suburban growth.[10] Linking growth machine networks to their spatial outcomes, Ted Muller has theorized metropolitan development as a "localized production system" that produces, among other things, manufacturing districts, housing, and infrastructure.[11] Robert Lewis identifies four key parts of the system—"production chains, the division of labor, external economies, and locational assets," the last of which include "a rich set of place-based inter-firm and institutional relationships" that tie together dispersed factory districts.[12] This framework describes mainly the dynamics within regions, though it also helps explain external relationships, for example Chicago's or Los Angeles's position in the world economy.[13]

Charting the relationships between individuals, firms, and institutions enables us to capture the ways in which diverse sorts of networks—of production, labor, infrastructure, and other social and spatial organization—overlap and shape one another. Engineers like the Sellerses tied together the four parts of Lewis's system in their various networks. In some of their projects they explicitly sought to define the physical and economic relationships between regions—what economic geographers call "systems of cities," though this term applies just as well to the systems they built within regions.[14]

The social, technological, and business networks of the Sellerses exerted both centrifugal and centripetal forces on cities and their economic activity. They fostered an intense *concentration* of information, skills, and markets in particular places, creating the density of interaction and productive capacity that sustained regional economies. At the same time, the ties that family members formed with people, firms, and institutions in other parts of the United States and beyond built *expansive* networks that allowed their enterprises and their cities to gain power in the national and global economy. They invested considerable energy in institutionalizing these relationships as well as their influence within and between regions.

The Institutional Complex

Like networks, institutions are especially useful units of analysis for uncovering and conceptualizing big processes such as urban economic development. The constellation of technical, promotional, social, and political associations the Sellerses joined and led illuminates the evolving structure and focus of economic development institutions. The diversity of their institutional affiliations and projects offers an opportunity to conceptualize a regional "complex" of institutions concerned with different sides of regional or national development.

Studying institutions lets us trace the ways in which people attempted to organize social and economic change. Economic historians of the last generation have tended to view institutions as the drivers and coordinators of economic change, after a generation or more in which most saw technology as the chief cause.[15] Social historians have researched institutions as a way to understand social movements, class, and the relationship between business and the state.[16] Within urban and planning history, public and private institutions are central to investigation of such topics as housing, environmental planning, urban renewal, and regional development.[17] Examining a complex of institutions can illuminate the contours of an entire sector—whether the chemical or social service sector—and even the metropolitan economy at large.

Following people across a set of organizations can capture much of the dense webs of overlapping networks. Institutions in themselves are networks of the people who join or frequent them, affording structured forums for individuals with similar interests to communicate and collaborate in furthering common goals. Their membership and its actions divulge a great deal about how different interests align to achieve specific aims.

The power of this mode of inquiry—follow the people and their projects, and by extension follow the money—is amplified by examining not just single institutions but a metropolitan complex. It can shed much light on their collective impacts and implications for urban life.

The Sellerses' institutional pursuits articulated directly and indirectly with the work of their firms, a common pattern for people engaged in many sorts of organizations. Their companies' product testing, marketing, and other projects with universities and technical societies furthered their business objectives. These and other activities tied innovation in business and technology to regional economic development. Ethical commitments figured more prominently in their participation in antislavery and other social reform movements. However, these activities often had more than a little to do with their economic concerns. Participation in many institutions, including those through which capital built power over labor, enabled the Sellerses and their associates to foster the social and environmental conditions supporting industrial growth, which constitutes a large part of economic development still today.

The varying strength of institutions and institutional complexes has significant consequences for people's ability to promote prosperity of cities and regions. Urbanist Jane Jacobs and business historian Philip Scranton have argued that the fortunes of metropolitan economies depend on their ability to build strong networks of institutions and firms, and on the level of interaction between them.[18] Certain members of the Sellers family, especially Nathan and later William, gained positions in especially influential institutions that profoundly affected the evolution of engineering, planning, and city building. Other institutions and projects in which they invested failed, underscoring some of the conflicts, limits, and unrealized visions of the Sellerses and fellow capitalists. Over the long term, their story illustrates how the fluctuating health of Philadelphia's complex of public and private economic development institutions was a determining factor in its growth and decline as an industrial center.

Metropolitan Geography and Material Life

The spatial and material interventions of engineers and their allies likewise played a defining role in urban economic development. In planning and building roads, canals, railroads, factory districts, steel buildings, electric power systems, and real estate ventures, the Sellerses strove to impress their own vision or order on the economy and society.[19] As much as it was the command and control center for industrialization, the metropolis was also an instrument of economic development, a tool in capitalists' efforts to control their own city's growth and to connect and compete with other regions.

The built environment and social geography of cities are among the most enduring and influential products of industrialization. Most of the major material transformations—whether coal distribution, telegraph lines, company towns, park systems, or elite suburbs—constituted networks and institutions in themselves. Taken together at particular points in time, these infrastructure and real estate projects reflected evolving paradigms of economic development, from the Enlightenment science Nathan and his father practiced in the eighteenth century to the early Progressive urbanism of William and Coleman in the decades after the Civil War.

Within the larger social and economic systems they inhabited, mechanics and engineers like the Sellerses made choices about where to invest—in what cities, sectors, and projects—that both reflected and shaped the geography of opportunity in industrial America. With other capitalists, they logically sought to benefit from and at the same time enhance their city's place in the larger colonial, national, and global economies. When they moved between cities and regions, as mechanics and engineers often did, their experiences revealed some of the sharpest tensions and contradictions in the uneven urban and economic geography of industrial America. Ultimately, as their particular stories intersected larger patterns and processes, the Sellerses and their contemporaries show how people's decisions and actions have shaped the fortunes of cities and their place in the wider world.[20]

CHAPTER 1

※

MANUFACTURING METROPOLITAN DEVELOPMENT

In 1790, George Washington moved into his presidential mansion at Sixth and High Streets, as Philadelphia became the nation's capital. A grand brick home paid for by Robert Morris, the financier of the Revolution, the mansion was down the street from the region's chief merchant countinghouses and a block from the State House. Yet these individuals and institutions represented the city's fleeting claims to political and mercantile power. A decade later, President John Adams and Congress decamped to Washington, and Pennsylvania's capital had left for the country town of Lancaster. Morris sat in debtors' prison, and New York would soon surpass Philadelphia in finance and trade.

Even as George Washington arrived, the city's economic future could be found across High Street from the president's house, at Nathan Sellers's home and wire workshop and in the shops of nearby metal manufacturers. This community of mechanics was the foremost center of metallurgical science and high-skilled production in the Americas. They clustered around the U.S. Mint on Seventh Street, whose director David Rittenhouse called on Nathan and his neighbors for their equipment and expertise. These men made industry a driver of urban and economic development.

Sellers's products, which found customers throughout the early United States, supported the growth of a diverse economy and of the metropolitan environment itself. In the workshop on the ground floor of his house and in the alleyway in back, Nathan and his brother David made wire screens and molds for millers of grain, paper, lumber, and gunpowder, as well as for cotton and wool spinners. They also supplied screens for bricklayers, plasterers, windows, and skylights.[1] Type founders, engravers, jewelers, and early steam engine makers employed technologies pioneered by Nathan. Other customers included printers, merchants, government offices, literary societies, and the University of Pennsylvania's medical school—in short, a broad cross section of the city's economy.[2]

The Sellerses played similarly diverse roles in public life. Nathan and his wife, Elizabeth, had moved into the city in 1781, the year combat ended in the War of Independence and their son Coleman was born. Over the next two decades, Nathan helped lead the Pennsylvania Society for Encouragement of Manufactures and the Useful Arts, and he joined the Philadelphia Society for Promoting Agriculture, the Society for the Institution and Support of Sunday Schools, and the Pennsylvania Society for Promoting the Abolition of Slavery.[3] From 1806 to 1812, he sat on the city council, arranging financing and providing technical oversight for the city's street, water, and lighting systems.[4] For some of these institutions, and in private practice as a surveyor, he plotted roads, canals, and the subdivision of land across the region.

Producers like the Sellerses made up a distinct class of skilled manufacturers who helped establish the social as well as material conditions to support urban and industrial development. As Escol Sellers later remembered, his grandfather Nathan "made no pretensions to be what is called a society man." Nathan referred to himself, somewhat self-effacingly, as a "tinker," a man who worked with his hands. He and fellow manufacturers allied with other sorts of capitalists to promote what they called "improvement."[5] This broad term referred to an Enlightenment vision of social and material progress, and in practical terms meant investments in research, technology, metropolitan infrastructure, and social reform to foster a growing and well-regulated, moral economy. The specific focus, meanings, and geography of improvement changed in the context and in the service of first colonization, then nation building, and in the early nineteenth century the spread of factory and wage labor. The Sellerses and their contemporaries took advantage of political and economic restructuring between these eras, and over the long term they crafted distinctly industrial modes of metropolitan development.

By the time of his death in 1830, Nathan Sellers's Philadelphia was a city of steam engines, textile mills, and infrastructure to support them, and his grandsons would soon make some of its first locomotives. It was also a place with a robust complex of institutions grappling with the challenges and opportunities posed by industrialization. These patterns were not just a product of the post-Revolutionary era, but rather had their roots

Figure 2. Charles W. Peale sketch portrait of Nathan Sellers at Mill Bank, c. 1810. An old family story has it that Nathan considered sitting for a portrait to be vain, so he agreed to sit for just twenty minutes for this "sketch portrait." Frick Art Reference Library.

in the region's colonization. Since the late seventeenth century, Nathan's ancestors had been milling, surveying roads and canal routes, and forming institutions concerned with promoting, regulating, and otherwise tinkering with metropolitan economic development. Their lives and work reveal the deeper origins of the region's industrialization, the evolution of its institutional complex, and the overlapping networks that structured growth. They render a portrait of early American capitalism, particularly in Philadelphia and other northern cities, that has much more to do with industry than most history books suggest.

The Industrial Colony

In 1681, Nathan's great-grandfather Samuel Sellers and his brother George left their hometown of Derby in the North Midlands of England for the new colony founded by fellow Quaker William Penn. During the next century, the Midlands would give rise to industrial cities such as Manchester, Sheffield, and Leeds.[6] But long before the great textile and steel mills, families like the Sellerses had developed milling and metalworking sectors, building corn and paper mills, iron forges, and nail manufactories. Samuel was trained as a weaver of cloth and wire, while George was a cooper.[7] In crossing the Atlantic, they participated in a vital transfer of industrial skills and knowledge from Europe to America.

British colonization in the Americas is rarely cast as an industrial project, except in providing raw materials and markets for English manufactures, but the experiences of the Sellerses and of Philadelphia show that it sometimes was. The mills of Manchester and the distilleries and shipyards of Liverpool surely grew on a steady diet of Carolina cotton, West Indies sugar, and New England timber. Yet some colonies were planned as more than extractive ventures. In Pennsylvania, William Penn and his investors envisioned industries that would create more complex and valuable products.[8] Migrants like the Sellerses were vital to this plan. They developed the agricultural and craft sectors that gave Philadelphia a meaningful place in the Atlantic economy and that put it on the path to industrialization.

As in other English colonies, the process of settlement meant creating urban, agricultural, and institutional geographies in which northern European farmers and craftspeople could replicate much of the economy and society of the places from which they came. Together with a group of fellow Quakers from Derby, Samuel and George settled just outside of Philadelphia, in a place they called Darby. The brothers slept in a cave or dugout on the side of a hill during their first winter there, but soon built a house and cleared land for their farm.[9] With their neighbors, they established a Quaker meeting. The first minute in its books, dated May 2, 1684, recorded that "Samuel Sellers and Anna Gibens of Darby, declared their intentions of taking each other in Marriage, it being [the] first tyme" for both.[10] George died when he was thrown from his horse two years later, but Samuel and Anna continued farming and weaving. Anna served as an overseer of the meeting, and Samuel helped to plan and police the township's settlement through his service as a juryman, supervisor, constable, and fence viewer defining property lines.[11]

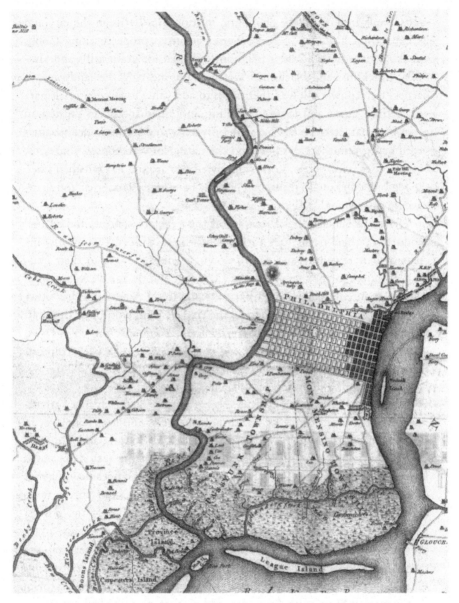

Figure 3. Detail of Scull and Heap's map of the Philadelphia region, c. 1777, showing the built-up city by the port with roads radiating out to towns, estates, and mills along the region's waterways. N. Scull and G. Heap, L. Hebert sculpt., *A map of Philadelphia and parts adjacent: with a perspective view of the State-House* (Washington, DC: Library of Congress, Geography and Map Division, image ID: G3824.P5 1752.S3 Vault).

Part of a constellation of satellite towns around Philadelphia, Darby and its residents occupied an important position from which farmers and craftspeople influenced the economic relationships between the city and its expanding hinterland. The Sellerses acquired property along Cobbs Creek, which they and their neighbors adapted for grain mills and sawmills, limekilns, and blacksmith's forges. Engineering the landscape to harness waterpower allowed their mills to add value to raw materials harvested from farms and forests.[12] Milling and allied industries created an economy characterized by increasing diversity and technological sophistication. Beyond the city's port and the cluster of warehouses, shipyards, coffeehouses, workshops, stores, and dwellings around it, craftsmen lined the region's creeks with mills, butcheries, chandleries, tanneries, and soapmaking operations.[13]

The early Sellerses lived mainly off the vegetables, fruit, and animal products of their farms, but each generation expanded their mills, increased their revenues, and passed down specialized skills to their children. Samuel's son Samuel Jr. gained a wide reputation as a weaver of wool and was reputedly the first weaver of metal wire in America, as his father apparently did no public business in the trade.[14] Samuel Jr. plied this craft in secrecy on the top floor of the family house, using machines of his own invention.[15] On a small scale, he sold woolen products to keep colonists warm and wire screens to help grist (grain) mills, bakeries, and gunpowder makers sift the fine particles of their business.

By the mid-eighteenth century, the growth of Pennsylvania agriculture and processing, combined with poor harvests and food shortages in Europe, made the region the breadbasket of the Atlantic and Philadelphia the foremost city in North America. The milling sector expanded its capacity thanks in part to Samuel Jr.'s son John, who became a supplier of equipment to other manufacturers. In 1763, he affixed wire screens and sieves to a water-powered reel in his mill, automating the process of removing the chaff from raw wheat, flaxseed, and other grains, a task previously done by hand.[16] He sold bolts (wire cloth), rolling screens, and sieves up and down the East Coast. In an ad in Ben Franklin's *Pennsylvania Gazette* in 1767 he claimed to be "the original inventor and institutor of that branch of business in America," adding that, since he had not taught his art to others, he was "in all probability . . . the best master of the work." He asserted that he had made all the wire bolts in use in Philadelphia and New York as well as a good number of the screens used for automatically cleaning wheat anywhere in the colonies.[17] John also added ninety acres to his father's property. He built a sawmill in the 1770s where two of his sons, David and Nathan, helped set up mechanical rippers, a draw bench for wire working, and a linseed oil press.[18]

In a deliberate strategy, John and his wife, Ann, took a diversified approach in the education of their five sons who reached adulthood. Nathan, the oldest, was indentured for a term of eleven months to train as a scrivener, or notary. He learned to draw up legal documents such as deeds, indentures, legislation, and wills, and to compound black and red inks in an era when ready-made ink was rare, especially in the colonies.[19] He subsequently served as a clerk for the Supreme Executive Council of the province and worked for attorneys in Philadelphia. In the 1770s, however, he interrupted this path to becoming a lawyer. Instead, he followed his father as the region's premier wire weaver and a leading surveyor.

Nathan's younger brothers Samuel and George trained with their father. They adapted milling technologies to the farm, which soon included large horse and cow stables where drinking water was conveyed automatically from the creek by a system of races and troughs adapted from milling techniques. David apprenticed with a trunk and hat maker in Philadelphia and John II at a tannery owned by their father.[20] The brothers' training mirrored and ultimately furthered the economic diversification of the region.

Like other English colonists in Philadelphia, the Sellerses subscribed to an Enlightenment worldview that held the human condition could and should be advanced through the rational observation and "improvement" of natural and technological systems.[21] Nathan filled his diaries and notebooks with sketches of milling equipment and machines, accompanied by long columns of numbers recording his observation and calculation of their performance under varying conditions.[22] Even after he "retired" because of poor health, he remained active in business and in tinkering. "I doubt whether it be right for any human being who has bodily strength, to excuse themselves from bodily employment, from the performance of uses," he wrote in his diary. "The human family must live by labour, and therefore it seems a duty, and cannot be dishonorable as many erroneously deem it."[23] Though he was often oblique in his written critiques, Nathan's efforts to distinguish himself from the region's merchants, together with his choices of which civic institutions to frequent, suggest his vision of improvement diverged from some of his contemporaries in his greater commitment to industry. This meant not just manufacturing, but in a larger moral as well as material sense it referred to the creation of economic value through hard work by free labor.

In the eighteenth century, improvement encompassed a broad suite of economic, technological, and social development interventions at different scales, from small adjustments making machines more efficient to extending metropolitan infrastructure and establishing institutions of social reform. As a family of surveyors, politicians, and manufacturers, the Sellerses engaged in multiple strategies of improvement. Nathan's father, John, served as constable of Darby Township, overseer of the poor, and for five terms in the Pennsylvania Assembly. In Darby and for the assembly, he surveyed private landholdings and public roads. With his assembly colleague John Lukens he surveyed land and water routes into the province's interior.[24] By fixing boundaries and determining major routes of transportation and communication, they replaced Indian trails and hunting grounds with a regulated landscape of private property and linked Philadelphia with networks of towns in its hinterland.[25]

Manufacturers like the Sellerses allied with the region's merchants to form a broader—though still small and often intimate—economic elite. Three of the brothers, Nathan, David, and John II, took parallel paths in marriage, wedding three sisters, Elizabeth, Rachel, and Mary, daughters of Philadelphia merchant Joseph Coleman. The intermarriage of leading manufacturers and merchant families would become a common pattern, solidifying the ties between different capitalists. But marriage was just one of the institutions that helped to organize economic integration and growth.

As the city expanded in the mid-eighteenth century, the Sellerses and fellow Pennsylvanians established a complex of institutions to increase and diversify agriculture, industry, and trade. In 1768 John became a member of the American Philosophical Society, one of a set of institutions established by Ben Franklin. Others included the Pennsylvania Hospital, to which John contributed ten pounds, as well as the Library Company and the Academy and College of Philadelphia (later the University of Pennsylvania). These institutions cultivated the specialized knowledge necessary to build a more complex economy and gave the region's elite new platforms from which to shape public affairs.[26]

The Philosophical Society quickly became the foremost center of American Enlightenment science. Its members formed a close-knit community of scientists, merchants, and manufacturers, many of whom were also active in politics. They included college provost William Smith, metallurgist and astronomer David Rittenhouse, as well as mathematicians, "mechanicians," surveyors, botanists, merchants, judges, architects, physicians, and printers. Early corresponding members hailed from nearly all thirteen colonies, the Caribbean, England, Edinburgh, Heidelberg, Stockholm, and Paris.[27] These connections to other regions aided the dissemination of knowledge, funneling news of the latest discoveries throughout the Atlantic world to Philadelphia.

A year after John joined the society, a Quaker miller, politician, mathematician, and astronomer named William Poole also became a member. Poole operated mills on Brandywine Creek above Wilmington, Delaware, where he served as town clerk. The two became fast friends. John's grandson John III would later marry Poole's granddaughter Elizabeth, linking the Sellerses to Delaware's community of manufacturers in a pattern of intermarriage that united the Quaker elite of the larger region. At the Philosophical Society, Poole and Sellers used their astronomical and surveying skills to help integrate Philadelphia into the scientific community of the colonial Atlantic.[28]

In 1769 the society launched its first large-scale research project, to record the transit of Venus, the passing of the planet between the earth and the sun, which occurs roughly once every hundred years. In petitioning the state legislature for funding, members stressed that their efforts were "calculated to promote the public good." Of especially "great consequence to these young countries" was their promotion of "every domestic improvement . . . that may help either to save or acquire wealth."[29] In its promise to advance navigation at sea and on land with new astronomical data, this research was a project of economic development, not an isolated science experiment.

The society established a pattern of collaborative research that tied scientists to manufacturers and merchants who applied research-based knowledge in business. The society paid fifty shillings to John Sellers for "a curious levelling Instrument of a new construction" that he invented for the project.[30] Poole watched the transit from Wilmington, while Sellers, Rittenhouse, and Lukens set up their telescope twenty miles northwest of Philadelphia, and other members recorded the event from points as far west as Fort Pitt (modern-day Pittsburgh), deep in the interior of the province. The society published the results in its journal, which it disseminated to institutions of science in the commercial and cultural centers of Europe and the Americas. This project literally and figuratively put Philadelphia and its scientific community on the map.

To further this sort of interaction and collaboration in different sectors, the society established various committees. Some organized around specific professions and branches of science, such as natural history; merchants; mechanics and architecture; and a medical committee that examined the purity of the city's drinking water. The Committee on Silk Culture studied European practices and promoted the development of that industry in North America. Sellers and other surveyors formed a Canal Committee to chart the most effective way of linking the Delaware River and Chesapeake Bay, confronting Philadelphia's growing competition with Baltimore for trade.[31]

By the time of the American Revolution, the Sellerses and other colonists had built a diverse regional economy with robust processing industries that positioned the region as an important center in the Atlantic economy. They had formed distinct classes of manufacturers and merchants who united in a broader elite to control much of the region's development. To regulate and promote their visions of improvement and public good, the Sellerses and their allies participated in government and built a complex of private institutions. Philadelphia was not yet a place of railroads and coal-fired factories, images that dominate popular conceptions of an industrial city. But its mills and networks of capitalists put it on the path to industrialization. This regional economy and institutional infrastructure also made Philadelphia the political capital of the thirteen colonies and the early United States.

Nation Building and Metropolis Building

The Sellerses played an important role in manufacturing the American Revolution, and their public lives illustrate how they and fellow Philadelphia capitalists organized economic development in the early Republic. In the decades following the war, the Sellerses helped fashion a new, more specialized complex of institutions dedicated to promoting metropolitan growth and grappling with its social consequences. This was simultaneously a regional and national project. Philadelphians struggled to retain the city's position as the economic capital of the United States. Though this proved beyond them, the networks of institutions and the infrastructure they built did enable the Sellerses and fellow manufacturers to sustain their influence upon economic and urban development.

Manufacturers made essential, if often overlooked, contributions to the War of Independence. During the war, Nathan joined his local militia, becoming a "fighting Quaker," though he saw no combat. Instead, the Continental Congress recalled him from his regiment to make screens and molds for production of the paper money that Congress used to pay for the war. At their sawmill, Nathan and his father and brothers fashioned tent poles for the army and made saltpeter for nearby gunpowder mills. Supplying these materials was not only a familiar economic role for the Sellerses but was also a logical outgrowth of their earlier collaboration with some of the manufacturers and merchants supporting the American side.

Nathan's father, John, participated early and actively in the economic protest that fomented revolution in and around Philadelphia. The region's diverse economy and society created broad constituencies for opposition to Britain's restrictive policies, including among producers who desired greater control over local economic development. John joined the Chester County Committee of Correspondence and attended the Boston Port Bill Committee, where Pennsylvanians adopted resolutions to boycott British imports. In the summer of 1774, his neighbors appointed him their deputy to the Provincial Convention.[32]

When war broke out, the Sellerses began supplying the rebel army with a range of products and services. Nathan's diary for the year 1776 records the manufacture of brushes and priming wires—pointed wires used to pierce paper packets of gunpowder before firing a musket.[33] In March, the Continental Congress engaged Nathan, his brother Samuel, and their father John to sign its new currency. In the absence of a government printing office, Congress employed the signatures of well-known, trusted citizens to validate each bill.[34] One day in July, Nathan recorded signing some forty-eight hundred bills in a single day.[35] Beyond his stamina, these activities reflected the Sellerses' central position within key manufacturing sectors as well as their alliance with a set of merchants and politicians within and around the Federalist elite of the region and emerging nation.

In August 1776, Nathan's reserve militia unit was called to northern New Jersey. He quickly drafted a will for his commanding officer. They got as far as Newark before Samuel and John arrived with the order from Congress: "A petition from sundry paper makers" had been "presented to Congress and read praying that Nathan Sellers . . . may be ordered to return home, to make and prepare suitable moulds, washers & utensils for carrying on the paper manufactory."[36] Like other supplies during wartime, high-quality paper was in short supply, as was paper in general, forcing people to tear pages from account books and diaries to write letters. Moreover, currency was not like other sorts of paper. It needed to be made in a way that discouraged false reproduction, and Nathan had surpassed his father as probably the best maker of fine papermaking apparatus in North America. The following April, the Treasury Board called him to the temporary Continental capital at York to make more molds for paper money.[37]

The rebel government also employed Nathan's surveying skills. In the summer of 1777, he aided preparations to defend the Delaware River, surveying a swath four miles wide and thirty miles long from Philadelphia to the Christiana River at Wilmington. At the instructions of the Pennsylvania Council of Safety, he paid particular attention to "the several places where an enemy may land, and the kind of ground adjoining, whether marshy, hilly, open, or covered with woods, and where there are several heights near each other, remarking their altitudes and distances apart." He documented the area's streams "as high up as the tide flows, and the places where they may be forded or passed by bridges—where there are swamps near the river, or roads—their kinds and sizes."[38] This knowledge of the region's waterways would facilitate his later development of mills.

Nathan's service in the militia and John's manufacture of supplies for the army caused the Darby meeting to disown them, two of roughly one thousand fighting Quakers who faced expulsion in Pennsylvania. Samuel walked a finer line, endorsing the Continental currency but avoiding activities that deviated from Quakers' dedication to pacifism. John appealed his excommunication yet failed to appear at his defense.[39] He would later reconcile and return to the meeting. Nathan joined the Swedenborgian faith, which advocated a similar socially progressive doctrine. He remained critical of religious organizations, writing later in life, "I think it very questionable whether any Human Authority, whether any man, or association or combination of men should be allowed or can have the right . . . to rule over the consciences of others."[40]

Despite this religious conflict and a raid on their Darby farms by Hessian soldiers, the Sellerses' losses in the Revolution were minimal, and their gains amounted to much more than their payments for equipment and provisions. In the postwar years, capitalists allied with the American side enjoyed an opportunity to establish a lasting, central position for their firms and their city in the national economy. The relationships between men who had served the Continental side, especially as army officers, contractors, and politicians, largely defined who was "connected" in the postwar economy. This mattered nowhere more than Philadelphia, whose merchants and manufacturers played large roles in financing and provisioning the war, and where the federal financial institutions would be established.

Separation from Britain enabled Americans to make their own financial system, patent system, and other institutions to regulate and grow the economy. Yet the end of the war brought new challenges to Philadelphia's central place in the American economy. As the city lost the state capital in 1799 and national capital in 1800, and as New York and Baltimore increasingly threatened its commerce, the stakes rose in the drive to capture markets in the North American interior and overseas. Moreover, across the postwar and early national period the rural majority and its political representatives, most famously Thomas Jefferson and Andrew Jackson, sought to limit the power and influence of urban capital in defining the economic geography and development of states old and new.

Many of the strategies for growth remained the same as before the Revolution. Nathan maintained an active surveying practice after the war, laying out roads and landholdings.[41] In 1784, the Pennsylvania Assembly engaged him and David Rittenhouse to revisit the possibility of a canal to link the Schuylkill and Susquehanna Rivers, to keep Baltimore from draining the produce of central and western Pennsylvania.[42] John Sellers likewise continued in public life in the 1780s, on a commission to build a new county courthouse and jail, as township supervisor, and as a representative for Delaware County (subdivided from Chester County) at the state constitutional convention in 1789.[43] Elected to the state senate the following year, he was subsequently appointed by Governor Thomas Mifflin as an associate justice of the county courts, but he chose to remain in the senate to push pending legislation for "Useful Improvements."[44] His main agenda centered on "more easy and speedy Communication between [the state's] remotest Parts, by the Improvement of its Roads and navigable Waters."[45] He also joined Robert Morris, David Rittenhouse, and Assistant Secretary of the Treasury Tench Coxe in the Pennsylvania Society for the Improvement of Roads and Inland Navigation, one of several promotional associations behind a wave of transportation investments in the 1790s.[46]

Although "internal improvements" would later become synonymous with just transportation investments, in the 1790s and early 1800s the term referred to a wider range of institutions and infrastructure established to expand, diversify, and integrate the nation's markets and communities. One thing that distinguished this era from pre-Revolutionary improvement was its more specialized complex of institutions. These included institutions of the new financial sector, those promoting development of particular sectors, education, and social reform aimed at an increasing class of laborers living in poverty. Through these forums, the region's capitalists, especially its Federalist elite, sought to retain control of the region and its economy as both became increasingly diverse and diffuse.[47]

The financial sector offered more than investment opportunities for the Sellerses and fellow manufacturers. It also promised to create a more liquid economy, funneling capital into large-scale infrastructure and a set of mostly well-connected enterprises.[48] In 1791, Nathan and David were among the early subscribers to stock in the Bank of the United States, which, despite its position as the national bank, mainly invested in Pennsylvania. When the Bank of Delaware County formed in 1814, Nathan and his brother John bought stock and regularly attended the election of its directors.[49] Nathan would amass tens of thousands of dollars in stock and bonds of the City of Philadelphia, the Bank of Pennsylvania, and turnpike and canal companies, all concerns explicitly devoted to regional economic development.[50]

The Sellerses continued to engage in a variety of interrelated scientific research, business, infrastructure, and educational ventures. It is unclear whether John remained involved in the Philosophical Society, but Nathan helped lead the Pennsylvania Society for Encouragement of Manufactures and the Useful Arts. In 1785, he and his father joined the Philadelphia Society for Promoting Agriculture, a new body made up not so much of small farmers as elite men with broad interests in economic development. At a meeting the next year, John presented a model for a permanent bridge across the Schuylkill River to link the city with the main roads to West Chester, Lancaster, and Baltimore. His experiments with gypsum, which he ground at one of his mills to rejuvenate the soils of his century-old farm, were of particular interest to farmers and chemists seeking new fertilizers to boost productivity.[51] As frontier farms increasingly supplied staple grains for eastern markets, this and other improvements tracked and tested by the society and displayed at its fairs aided a local transition to higher-priced dairy, eggs, and vegetables. This, in turn, increased farmers' demand for manufactures. The society's efforts led to the University of Pennsylvania's establishment of a Faculty of Natural Sciences and Rural Economy, though its plans for a veterinary school, pattern farm, and botanical garden saw only partial implementation.[52]

In 1794, Philosophical Society member Charles Willson Peale, whose daughter Sophonisba would marry Nathan's son Coleman, established a museum of natural history.[53] This was a new sort of institution, open to the middle-class public, unlike associations that invited only select "leading men." Dedicated to "diffusion of knowledge" through "rational amusement," the museum offered entertaining lectures and science demonstrations. Its permanent collection later included plants, animals, and Indian objects brought back from the West by Lewis and Clark. Across the 1790s, Peale sought to persuade the federal government to make it a national institution.[54]

For the Sellers children and other Philadelphians, it became a valuable educational institution. Young Escol found great excitement when his uncles gave him a part in the electrical show, in which a charged wire and a cork blown out of a bottle of gas simulated lightning and thunder, destroying a miniature model house.[55] Nathan half-jokingly called Peale a "show man," in contrast to his own identity as a "tinker."[56] But like the Sellerses' workshops and mills, the museum allowed their grandchildren to explore science through artifacts and machines. It represented an important foray into scientific education for the middle class.

Nathan's most public role in Philadelphia's economic development was his service on the city council from 1806 to 1812. Municipal and state government complemented the promotional societies whose members, like Nathan and his father, often entered politics. Through the public sector, these individuals helped craft a well-regulated economy and society in the early Republic, enacting legislation that covered exchange, public health, safety, and welfare.[57] Financing and technical oversight of urban and economic development was also much of the job.

In an era when the city council was responsible for municipal planning, finance, and public works, Nathan's experience rendered him especially fit to serve. Employing his skills as a scrivener and his knowledge of surveying, he drafted ordinances that governed the layout and leveling of new streets as well as assessment of real estate taxes. His expertise as a mechanic must have informed his service on the mayor's committee on weights and measures, which set standards for the city's port and market activity. He filled the notebook he used in council with sketches and calculations of cost estimates as well as notation of bills introduced and passed.[58]

Building and maintaining urban infrastructure was perhaps his biggest job on the council. As a member of the standing committee on ways and means, Nathan managed the financing of the waterworks as well as the Chestnut Street municipal wharf through bonds and loans from the Bank of North America and the Philadelphia Bank. In his personal investment portfolio, he acquired $9,000 (roughly equivalent to $159,000 in the year 2010) of this same city debt.[59] In council, he also oversaw the yearly budget for paving, "cleansing the City generally," "repairing & cleaning docks & sewers," "lighting & watching the City," "Watering the City by means of pumps & wells," and salaries for the small bureaucracy and constabulary force.[60] This infrastructure was critical to supporting rapid growth. Between 1810 and 1830, the county would expand from 110,000 inhabitants to 190,000.[61]

In 1812, Nathan studied options for repairing or replacing the waterworks, a plant with profound implications for public health in the city.[62] Philadelphia had been ravaged by yellow fever in the 1790s and experienced regular outbreaks of cholera, which contemporaries blamed on disease vectors transmitted through wells and streams often contaminated by slaughterhouses, tanneries, and other industries. He opposed the engineer Benjamin Latrobe's plan for a new waterworks since it would block High Street, the city's main thoroughfare, and, in Escol's words, because of "the absurdity of supplying a future great city from a wash basin in the dome of a Grecian Temple." Nathan also disputed the plan, since it would draw water from the Schuylkill River's tidal basin where human and animal waste washed up daily. Instead, he advocated a tunnel through which cleaner water from upstream could supply the city, and called for the city to purchase the hills along the river for future reservoirs. Latrobe's "temple" rose on High Street as planned, though two decades later an aged Nathan paid one of his last visits to Philadelphia to witness its demolition.[63]

The Sellerses, like other Quakers and Swedenborgians, sought to complement these material interventions in the health and safety of the city through a set of private social reform associations. Established in the decades after the Revolution, these institutions gave elite Philadelphians and their counterparts in other cities a range of opportunities to influence different segments of the growing working classes. Through their respective work, they distinguished and sought to tackle—albeit usually quite modestly and moderately—a set of social problems that the contemporary government, with its handful of constables and overseers of the poor, could not handle.

Nathan, his father, and brother Samuel joined organizations aimed at different groups of poor Philadelphians. The city's Humane Society, which Nathan joined in 1805, supported emergency response functions later performed by police. The Society for the Institution and Support of Sunday Schools paid private schoolmasters to teach reading and writing to poor children and adults on their day off, using the Bible as their main text. The all-white Pennsylvania Society for Promoting the Abolition of Slavery petitioned Congress and the state, assisted blacks in legal claims to freedom, and arranged indentures for children and employment for adults. It collaborated with the Free African Society, led by prominent blacks, in house visits promoting the work ethic and morality. Two years after Nathan joined, the Pennsylvania Society began a school for blacks.[64]

Civic pursuits such as these embodied mixed goals for capitalists like the Sellerses. These pursuits were often inspired by faith, and such affiliations are certainly consistent with Quaker and Swedenborgian commitments. The Sellerses and those like them were also surely driven by their need and desire for a well-regulated society in which the laboring classes would work hard and abstain from behaviors that threatened productivity, such as drinking or violence. This combination of motives points up how the Sellerses and many of their contemporaries continued to view improvement in more than simply economic terms.

Nathan's words and actions evince the tension between charity and manipulation that is common to social reform led by elites. In a love letter to his future bride Elizabeth in the winter of 1778–79, he expressed genuine compassion for the poor. Poor harvests and speculation by merchants in wartime drove up the price of flour and other staples. "The hard bound situation of Nature at this time calls for the Imagination to stalk over the face of the Earth and to peep into the Dwellings of its Inhabitants," he wrote. "Oh methinks, I see thousands shivering round a few dying Embers, with tattered garments and an uncheery look—No stores to Gladden their hearts—No Comfort—pinching Necessity all around—how anxious and unhappy the mind—poor Creatures." Faith, in his view, should inspire those more fortunate to act, as "there is a God who can open the hearts of these who have & to spare, and most Certainly it is their duty to lend a helping hand." More than charity, Nathan suggested all people should be content with "enough": "In these times of public Calamity the Earth teems forth no Overplus to the Inhabitants thereof—its produce is perhaps but sufficient to let all have the Necessaries—how Unjustifiable then the Wish to have a profusion."[65]

Yet participation in social reform institutions, coupled with some of Nathan's later business dealings, illustrates how elite concern for those less fortunate was inextricably tied to the regulation of their behavior. Sunday schools, for example, pushed a particular version of morality while also spreading basic reading and writing skills that were increasingly important for even low-paid laborers. In the 1790s when Nathan joined the Abolition Society, it intervened in the social and economic lives of the city's blacks as much as it advocated their liberation in the moderate mode one historian characterized as "Gradual abolitionism and a lawyerly chipping away at slavery's margins."[66]

Nathan's business relationships were sometimes quite paternalistic. Escol remembered a visit with his brother Charles to the family farm in Darby, where Nathan moved in 1817: "Grandfather came into the shop where we were tinkering at something and said to me, 'Escol, Joel Bonsall has gone into the house with butter and eggs. I want thee to go in and see what thy Grandmother buys and just what she pays.'" The boy reported the rate of nine cents for a dozen eggs and twenty cents per pound of butter. Nathan told him, "'Thee ought to know that hens can't afford to lay eggs at less than a cent apiece or cows to make butter at less than 25c [*sic*] per lb. and if Joel don't get that price he cannot pay me the interest let alone the principal of his mortgage.'" He handed Escol the difference and sent him after Joel, "whom he said I would find had gone no further than the spring house." On his return, Nathan explained, "'Thy Grandmother has a frugal mind and it makes her very happy when she makes a good bargain.'" Escol replied, "'She says she does not think it wrong to beat Joel down for he spends all his money for drink and the less he has the better for him.'"

Though Elizabeth appeared to be the manipulator here, Escol later learned from Joel himself that his grandfather had neglected to share the whole story. Joel explained that Nathan "had proposed to pay him a certain fixed price for butter, eggs and other farm products on condition that he was to keep for [Nathan] the difference between what Grandmother paid and the price fixed on, which he found paid his interest" on his farm for which Nathan held the mortgage. But Nathan got one more thing out of this deal. Joel "promised to quit drink and weekly hand to Grandfather in trust his drink savings and . . . soon his farm began to prosper" and he paid off his debt.[67]

From his public life to his personal business, Nathan sought to build the sort of moral economy and society that reflected his Swedenborgian sensibilities and the industry of his active hands and mind. His family found a community of common interest among other Philadelphia capitalists keen to maintain the power they gained through the Revolution. They employed government and charitable institutions to protect and promote the "public" economic interests defined by producers and their allies in the mercantile class. Through this institutional complex they helped plan, underwrite, and regulate metropolitan growth in the early years of the Republic. In the process, they established a pattern of economic development—including social reform—that would spur industrial growth.

Even in his retirement, Nathan remained passionate about internal improvements and his region's competitive position in the national economy. After examining plans for the Erie Canal, in early 1822 he penned a frantic warning to the state legislature. Calling the project "Monstrous! or rather Stupendous," he lamented, "as a Pennsylvanian, I am alarm'd with the magnitude of its consequences." He encouraged the state to "run a good deal in debt to make such facilities of transportation as would prevent the resources of it being run away with by other people, as well as to share in the transit of other peoples products to our own metropolis."[68]

Though the legislature followed this advice, Nathan's fear that New York capital would dominate the West and South came true.

Instead, Philadelphia's nineteenth-century prosperity was based on a different pattern of economic development. Its manufacturers, railroads, and financiers gained a sphere of influence concentrated among the hundreds of mining and factory towns dotting Pennsylvania, Ohio, and nearby states. Their social, economic, and infrastructure networks created a more intensive pattern through which the city became a central place in the nation's industrial heartland.[69] Boosters would dub it the Manchester of America, no longer the largest or most powerful city in the nation but its leading center of industry, with a highly diverse economy. While New York had more manufacturing jobs for most of the nineteenth century, on account of its larger population and economy, Philadelphia remained a more important center of engineering and innovation. The Sellerses did a great deal to foster this transition to a full-blown industrialization.

The Pillars of Industrialization

Nathan and his family played a large role in the four material and social transformations that historians generally agree define the first industrial revolution: mechanization, factory systems of production, wage labor, and mineral sources of energy.[70] In the early nineteenth century, he and his children focused their firms in the three major sectors that drove these changes: textiles, metalworking, and transportation. They developed cotton mills that helped turn the region's mills to textile production, the largest manufacturing sector on both sides of the Atlantic in the early and mid-nineteenth century. Nathan's son Coleman succeeded him in the metal business, turning it into a leading maker of papermaking machines, fire engines, and early locomotives in the 1820s and '30s. Both of these ventures relied heavily on social and business networks that reveal how the Sellerses developed new technologies and sectors.

The businesses Nathan and his family built in the early nineteenth century supported a wide range of industries and their networks. Nathan and David's wool-carding combs, hooks and eyes, and knitting needles fed the textile trade. Their sieves and screens remained necessary and pervasive in milling and the building trades. Their main specialty, however, was paper molds.[71]

All these products benefited from an annealing process Nathan developed, heating metal to a temperature at which it changed chemical structure. This preserved the metal's shine and prevented rust while making it more malleable and ductile.[72] His wire equipment thus became easier to bend into more diverse and refined products. It also held up better in use, an advantage that helped grow the Sellerses' reputation and markets.

As specialized metalworkers serving other producers, Nathan and David sold to a substantial market beyond the region, unlike most manufacturers in the early nineteenth-century city. Their customer base included most of the prominent early American papermakers, as well as wholesale brokers and shipping merchants who supplied the South and the Caribbean. They also sold to banks, institutions, and wealthy individuals who ordered paper made to particular specifications and with personalized watermarks.[73] This helped assert the identity of individuals, corporations, and institutions engaged in dispersed networks of printed communication, which together with the transactions and obligations recorded on this paper made the economy liquid.

Nathan and David grew their business by expanding networks of supply, production, and distribution. Their correspondence reveals a broad web of relationships with merchants and manufacturers. They bought brass wire from London, iron wire from a Pennsylvania forge, and tin from a local wholesaler. They contracted with carpenters and cabinetmakers to build the wooden frames for their molds.[74] Such import, export, and contracting activities continued to link—and blur the lines between—mechanics and merchants, and between mechanics and craftspeople, as wholesalers and subcontractors were motivated to encourage growth in other sectors.[75]

Relationships with colleagues in associations promoting manufacturing and internal improvements as well as other institutions aided the Sellers firm. Evidently Quaker networks still mattered for regulating business. During the War of 1812, when access to England was cut off and demand for American manufactures boomed, the merchant-banker Stephen Girard, who frequently shipped the firm's wire sieves and screens to the West Indies for use by sugar planters, offered to secure an order from London via Germany in neutral vessels. As recounted later by Escol, "The Quaker firm of Trotters, then in the metal importing business, got wind of its arrival, went to Girard and offered him" more than the Sellerses had. But "Uncle David took the matter up and presented the case to leading members of the Society of Friends, who so shamed the Trotters' they withdrew their offer with the lame excuse that they had not known it was a special order of N. & D. Sellers'."[76] Girard, however, assumed he could still get the higher price on the Philadelphia market, so he advertised the order for auction. At the sale, the only bidder was the Sellers firm. All others present abstained.

Even as Nathan and David built their downtown business, they maintained a wire workshop and mill in Upper Darby. They and their brothers played diverse roles in Delaware County milling, often influencing entire sectors. Their brother George operated the saw and paper mill built by their father. David's son David Jr. built a tilt-mill for hammering metal.[77] After a stint operating tanneries in the city, Nathan's brother John returned to Upper Darby and in 1804 inherited their father's flour mills.[78] He also built a new one, named Millbourne, which his son John III and grandson John Jr. would successively operate into the twentieth century. According to a company history of 1888, "The price of wheat in counties west of Philadelphia was regulated by this mill," a sign of the status and influence the Sellerses enjoyed in this sector in the early nineteenth century.[79]

In collaboration with their colleagues in the region's scientific institutions, the Sellerses enhanced their milling practice through the study of chemistry and physics coupled with careful observation of the mills' mechanical performance.[80] Nathan, his brother John, William Poole, and other millers in the family corresponded about their experiments with different wheel diameters, the number of cogs in each wheel, revolutions per minute, the amount of waterpower needed to run at various speeds, and the effects of each of these variables upon their products as well as the wear and tear on equipment.[81] John invested heavily in automating his flour mill.[82] Nathan submitted his drawings and model of a simplified gate to manage water flow to the Philosophical Society, where a committee vetted the invention, published it in the society's journal, and placed an announcement of the invention in the newspapers.[83]

Nathan and Elizabeth sent their son Coleman to Quaker school, though he gained as deliberate an education in the workshop and at family farms and mills. Relatives clearly encouraged him to tinker. As a teenager, he fashioned an alarm clock to wake him up in time for chores at his grandfather's farm, rigging up a bucket to catch water from the spillway of the nearest mill dam, so that the bucket would fill up gradually overnight and then spill out. Once full, the bucket would trip a lever that opened the sluices, emptying the dam. This rush of water turned a water wheel attached to a crank that tugged on a long wire Coleman attached to a cowbell by his bedroom window. But the first time he tried this system, a thunderstorm filled the dam by the middle of the night, at which point the water was released and the bell rang violently. According to family legend, the entire household was awakened, wondering how a cow had been stranded on the roof.[84]

Coleman succeeded his father in business, while Nathan continued tinkering and investing in new businesses and technologies after his retirement to Upper Darby in 1817. In 1820 he patented "a very useful and valuable Improvement in the cleaning sifting seperating [sic] and dressing Rice."[85] Two years later, at age seventy-one, he took up intensive study of trigonometry. Escol remembered, "One of Grandfather's fads or amusements" at his lumber mill "was the testing of the amount of shrinkage of timber cut at various seasons and the length of time in seasoning when shrinkage ceased."[86]

Still, Nathan complained in his diary of not feeling sufficiently "useful."[87] Escol recalled, "in physical condition Grandfather's decline was rapid and distressing" at this time, which he attributed to "the ruinous habit of blood letting by the advice of his friends," including the city's leading physician, Dr. Benjamin Rush. Trying to cope with bouts of vertigo, which they blamed on years of breathing fumes from annealing, Nathan gave up drinking entirely. But as Escol related, he remained an "inveterate smoker . . . being rarely seen from the time he got up until he went to bed at night without a cigar in his mouth."[88]

His health notwithstanding, Nathan maintained connections to fellow Philadelphia manufacturers, partly through Coleman, who ran the wire works with his cousins Samuel and James, whose father David had died in 1813. Coleman moved into the High Street house with his wife Sophonisba and their children. That same year, 1817, he formed a new partnership, with banknote engraver Jacob Perkins, to produce fire engines. They set up their workshop just around the corner in Mulberry Court, an alley that ran behind the High Street house. In 1829, the family would move around the corner to Sixth Street, where Coleman commissioned a new house with a warehouse that fronted on Mulberry Court.[89] This remained the center of the region's metalworking industry.

Coleman ended his partnership with Perkins quickly, dissolving it after Perkins's aggressive and unsuccessful attempt to outdo their neighbor Pat Lyon's fire engine, which Coleman had cautioned against and which caused a child spectator's foot to be impaled on an iron spike.[90] He formed a more successful venture in 1818, when the three Sellers cousins joined with

Abraham Pennock (who had married their uncle John's daughter Elizabeth in 1810) to make riveted leather hose for fire companies throughout the East and mailbags for the U.S. Postal Service.[91] The firm would go on to make fire engines and steam-powered papermaking machines in the 1820s, and later locomotives.

Perkins's rivalry with Lyon notwithstanding, the Sellers workshop was a magnet for a community of skilled mechanics who came to work, observe, and experiment. Although Coleman's family lived amid the homes and offices of the city's social elite, Escol asserted that neither his father "nor mother made any [pretensions] to the St. Peter's set or the Codfish aristocracy . . . High wine, sugar, rum and molasses with a spice of African blood," meaning he was not close with the city's most powerful merchants and slave owners. Instead, "it was among *producers* and *scientific* men he had some intimate and warm friends."[92] Those who regularly gathered in the front office on High Street included Lyon, machine builders Matthias Baldwin and Isaiah Lukens, U.S. Mint director Professor Robert Patterson and his successor Franklin Peale, and Dr. Thomas Jones, superintendent of the federal patent office in the late 1820s.[93]

The concentration of mechanics, scientists, and institutions in close proximity fostered networks of information exchange and production that benefited the Sellerses' business. Metalworkers in this district enjoyed symbiotic relationships with the nearby political, mercantile, and institutional core of the city. Like Nathan and David before them, Sellers & Pennock made equipment for the mint, a sign that they did some of the most specialized, precision metalwork in the nation.[94] Their shop remained a social center for mechanics and businessmen interested in metalwork and the diverse economic activities it enabled.

As Escol recalled, one man who frequented the shop was Commodore Murray of the Philadelphia Navy Yard. On a friendly wager of an oyster dinner, he challenged Jacob Perkins to devise a pump to bail boats at a certain rate. In attendance at the trial where Perkins demonstrated the pump he designed with leather buckets were two of Coleman's friends, wire manufacturer Josiah White and his partner Erskine Hazard, who "were engaged in their effort to float anthracite coal down the Lehigh and Delaware in small barges," Escol wrote. Perkins won the contest. "Mr. White was so well pleased with" the pump and buckets "that he adopted them for their barges, from which they found their way to the flat boats of the Susquehanna, and finally to . . . the Mississippi and its tributaries."[95] For the Sellerses and their colleagues, disseminating new technologies was an explicit goal that promised to grow their markets and access to resources. Their ties to federal institutions like the mint and the navy, and to builders of transportation and energy infrastructure, helped make this possible.

In their community of steam engine and machine makers, no resource would be more significant than the anthracite from eastern Pennsylvania that White and Hazard made effectively Philadelphia's coal.[96] The Sellerses had an important hand in this venture. After buying much of the anthracite territory in the Lehigh Valley, in 1818 they engaged David Sellers Jr. and a colleague to survey the area and provide an expert opinion of its development potential. By White's account, David Jr. and his partner "reported the Coal Mines inexhaustible but that the country was so Ruff between the [Delaware] River and the Mines, that they found it an easement to Run over Stumps to avoid Stones!"[97]

Because of the high costs of the transportation improvements necessary to bring the coal to market, White wrote, Nathan Sellers "recommended that we seperate the Navigation from the Coal & make 2 Seperate Cos. which alltho contrary to our notions of Right we Submitted to, for the sake of starting the work under his Recommendation, for he felt a freedom to Recommend the Navig[ation] on our plan."[98] White and Hazard believed Nathan's full endorsement would draw sufficient investment from the region's manufacturers, at a time when the region's bankers and brokers balked at underwriting them. Nathan invested $10,000 ($177,000 in year 2010) in the Lehigh Navigation Co. securities, and more later.[99] Through his promotion its initial stock offering was fully subscribed.

Anthracite ultimately replaced lower-grade bituminous coal and wood as the preferred energy source of heavy industry. Its higher heat values allowed steam engines to generate more power to drive machines. This energy revolution spurred the development of mechanized metropolitan infrastructure and enabled manufacturers to organize larger-scale, more regimented "factory systems" of production. Philadelphia capitalists' control of the anthracite fields would support a vast development of heavy manufacturing and make the city one of the world's largest coal exporters.

The machines the Sellerses made in the 1820s increasingly ran on coal, and they continued to serve the expansion of other manufacturing sectors and of cities and regions. Their fire engines addressed one of the foremost dangers of nineteenth-century urban life. Cities' wood buildings were threatened by conflagrations that regularly destroyed entire blocks and districts of homes and businesses. The firm also supported the development of public and elite private fire companies. Their engines' technological superiority over less efficient systems of buckets helped arm these customers in their struggle to usurp the business of firefighting from violent fire company gangs of working-class communities.

Sellers & Pennock were well positioned in the landscape of fire companies, for a time their most important customer base. James Sellers and Abraham Pennock were members of the Philadelphia Hose Company. Coleman's brother-in-law and sometimes employee Franklin Peale, who managed his father's museum on the upper floor of the State House a block away, set up a central "system of bell signals to indicate the direction of a fire," his nephew Escol remembered. When news of a fire reached the State House and Franklin rang the bell, "Then came the great rush, the madding races, the ringing of the bells on the engine and hose carriages, the bellowing 'Fire' through speaking trumpets that every fireman was armed with; it was pandemonium broken loose."[100]

In 1822, Sellers & Pennock combined a steam engine and hose carriage in the first working suction fire engine in the United States, named the Hydraulion. They used cylinders cast at John Wiltbank's workshop at 262 High Street, copper work supplied by Israel Morris's Third Street shop, pumping levers forged and fitted by eminent mechanic and Sellers family friend Oliver Evans's Mars Works, and woodwork fashioned by a carriage maker in West Philadelphia. The firm sold large versions of this engine to the fire companies of Philadelphia and Providence, Rhode Island, Washington, DC, and Alexandria, Virginia. Smaller towns and the University of Virginia ordered smaller versions. Since the Hydraulion was a new and complicated machine, members of the firm typically delivered and set up the finished product personally, training firefighters in its use and, in one small way, spreading mechanical skills throughout the East.[101]

The Sellerses' increasingly complex machines inspired new patterns of work. The specialized jobs in their workshop reflected the expansion of factory systems of production. With a growing division of labor, the shop employed roughly ten to fifteen men during the busier periods of the 1820s, as young men, including Coleman's sons, assisted the smiths and mechanics by pumping the bellows and running errands.[102]

These dynamics also produced a new geography of manufacturing. In 1824, the Sellerses left their tight quarters downtown and established a factory to make steam-powered sawmill and textile equipment, the most complex of their products, at Sixteenth and High Streets, ten blocks west. According to a later account by Coleman's son Charles, "Father fitted up the shop . . . and ran it for one year, then Cousin James took charge of it and got into such a mess that before his year was out he took sick and gave it up. Then father put me in charge of it to fill up the orders

Figure 4. Sellers & Pennock advertisement, showing the Hydraulion in use, c. 1818. Image: F8 33 2. American Philosophical Society, Philadelphia.

and work up the stock on hand and close it up" in 1827.[103] In the fallout from this failed venture, the Sellers cousins dissolved their Philadelphia partnership. David's sons James and Samuel continued to make wire equipment in the High Street shops. Coleman and his sons built a highly mechanized factory called Cardington in Upper Darby to manufacture carding combs and steam-powered papermaking machines, under the name Sellers & Sons.

As a firm on the leading edge of mechanical production, fashioning some of the most complex machinery of the age, Sellers & Sons drew customers and correspondents from London and Paris to Boston, Cleveland, Knoxville, and Charleston, South Carolina.[104] Escol claimed in his memoirs, "There was not a Paper Mill south of Mason and Dixon's line that we had not something to do with in the way of machinery," as the South lacked makers of sophisticated machines.[105]

While Coleman maintained a downtown office on Sixth Street that afforded access to colleagues, customers, and business and technical information, Cardington became one of the major industrial works of the East. It was in a class with Evans's Mars Works on Ninth Street, New York

City's Novelty Works, the Matteawan Co. and the West Point Foundry on the Hudson River, and the shops at Saco, Maine, and Lowell, Massachusetts. These facilities were organized around several departments: a machine shop with lathes and other tools; a blacksmith shop with welding capacity; a foundry where brass and iron castings were poured; a smith shop to roll, hammer, and slit wrought iron; a pattern room; a drafting room; and an assembly area where large machines and engines were put together.[106] They were at the forefront of factory systems of production in high-skilled manufacturing.

Cardington turned out a range of products, from paper molds to steam engines, catalyzing the development of various sectors. Even papermakers in Birmingham, England, one of the centers of the British industrial revolution, purchased Cardington strainers and machines.[107] Customers closer to home included other machine makers, paper mills, coal-mining companies, institutions such as Eastern State Penitentiary and Girard College in Philadelphia, and the branch U.S. Mints established by President Andrew Jackson in North Carolina and Georgia.[108]

Sellers & Sons fostered networks of production employing high- and low-skilled workers alike, mediating the development of important new sectors. Escol remembered that, for their papermaking machinery, "the different parts were so scattered about" among specialized metalworkers.[109] The cards produced at Cardington were much simpler products, but they enabled other networks in textile manufacturing. The cards were like iron brushes, used for separating raw cotton before spinning on a loom, the first step in the production process upon importing cotton from the South. After making and bending the wire teeth, the firm sent the teeth out to thousands of women and children working from their homes, who set them in the leather backing that they or other home-based workers held as they pulled apart the cotton. This enabled millers to spin the carded cotton into thread and weave it into cloth (as did home-based spinners and weavers, though they did a decreasing portion of this work).[110] These networks drew much of the population into the labor market—and increasingly into the mills themselves—of the textile districts that grew up across the region in the second quarter of the nineteenth century.[111]

While metalworkers drove mechanization and the adaptation of mineral energy, textiles led the way in spreading industrial wage labor among the working classes. With rapid adoption of the cotton gin, high tariffs on imports of British cloth, and expanding slave populations and cotton production in the South, the textile industry took off in New England and

machines mechanized both factory production and urban infrastructure. Their investments in anthracite coal helped Philadelphia secure a vital supply of energy for the rest of the nineteenth century. Their firms established factory systems of production at the high-skilled, specialized end of the labor market, while some of their products and properties structured opportunities for low-wage labor. In the process, the Sellerses and other manufacturers reshaped the geography and dynamics of work and class, defining the roles of different sorts of industrial workers. The institutions they deployed to promote economic development aided these shifts. However, their scientific, business, and social reform societies confronted only parts of the great social challenges presented by factory systems, wage labor, and rapid urbanization. They were not organized to engineer the sort of large-scale social project that would enable the explosion of the industrial workforce in the mid-nineteenth century, namely public education. To this end, the Sellerses and fellow industrialists therefore invested in yet another new set of institutions.

Education for Industrial Development

While the array of reform societies Nathan and his family joined after the Revolution addressed vital problems of public order, slavery, and education, they did not engage the great majority of Philadelphia's population. As wage labor began to replace the apprenticeship system among craftspeople, and as farm families and immigrants arrived in the mill districts, the region's capitalists initiated another transition in the complex of institutions that organized economic development. They invested in two new sorts of institutions in particular: those that cultivated specialized knowledge among mechanics and manufacturers, and those that spread basic education among broader publics, culminating in the formation of state-supervised school bureaucracies.

Sellers family members generally shared a worldview in which liberal education, free labor, and temperance were mutually reinforcing elements of a moral economy and society. They sought out like-minded Philadelphians to institutionalize these values. Nathan's nephew Samuel joined the Pennsylvania Society for the Promotion of Public Economy soon after its founding in 1817. Its broader mission was to promote the reform of poor relief by "ascertaining and pointing out" the "most profitable direction" for labor, inculcating in the poor the "prudent and judicious expenditure of money," "instructing the great mass of the community in the modes of economizing in their fuel and diet," and educating "the ignorant and the poor" in order "to strike at the root of poverty and vice."[118] Viewing intemperance and unemployment as the prime sources of these ills, it aimed to redefine the state's role in the life of the poor.[119] Samuel sat on the Committee on Vice and Immorality. Other committees addressed poor laws, elections, public prisons, public schools, and domestic economy. Their impacts in most of these realms were modest, though they won some changes, most notably in education.

In 1818, the society pushed through the legislature an "Act to provide for the education of children at public expense" within Philadelphia County. It mandated the erection of schoolhouses, hiring of teachers, and formation of a board of controllers. The act assigned "respectable taxable inhabitants . . . to superintend the schooling of the poor children" in different parts of the county, "and determine what children may fall within this description."[120] It did not result in universal schooling, and it would be almost two decades before a statewide public school act passed. However, the authority bestowed on property owners to define who was poor and should attend "common" schools was a significant step in reformers' efforts to regulate the industrial labor market.

Public education promised the people who controlled it greater influence over workers whose identity and roles in the economy would be increasingly defined through schooling. As historian of education Michael Katz points out, this "drive toward institutional innovation *preceded*"— and was a prerequisite to—the mid-nineteenth-century boom in industrial employment.[121] The basic reading, writing, and math skills taught to students who attended even just a few years of formal schooling served manufacturers' need for workers who could read written orders and make simple measurements and calculations.

In their educational investments, however, the Sellerses focused more attention on the needs of high-skilled manufacturing than on basic literacy and discipline of the common schools. This reflected the needs of their firms as well as the status of family members as workers. In the early nineteenth century, millers and other craftsmen increasingly defined themselves as "mechanics" (and ultimately mechanical engineers), a reference to the specialized knowledge they possessed.[122] Growing this class of workers required technical education and what today we call "knowledge management."

In 1820, Coleman and Samuel joined a large group of manufacturers and civic leaders in founding the Apprentice's Library of Philadelphia. Samuel served as its treasurer from 1826 to 1840. Unlike the Library

Company, which had become an elite organization with a closed membership, the Apprentice's Library offered free circulation and was open to all literate white workers. By 1830, it housed some six thousand volumes.[123]

Training for the region's top scientists also got a boost in the 1820s. The University of Pennsylvania increased its scientific faculty, courses, and laboratories in this decade.[124] Coleman's eldest son, Charles, would enroll in the medical department, training as a physician in addition to his education as a mechanic. In 1821, industrialists established the Philadelphia College of Pharmacy, which provided technical education and skilled labor for the region's robust chemical sector. The college initiated the *American Journal of Pharmacy* and published the first American *Pharmacopoeia*.

Three years later, a broad alliance of capitalists founded the most important forum for coordinating technical education, communication, and innovation, the Franklin Institute for the Promotion of the Mechanic Arts. Initiated by merchant Samuel Vaughan Merrick, who had taken over Jacob Perkins's bankrupt fire engine shop, with University of Pennsylvania chemistry professor William H. Keating, the institution attracted members ranging from gentleman farmers to stockbrokers. But its bylaws required that two-thirds of its board "shall be manufacturers or mechanics." The Sellerses were early members, and other machine builders who gathered at Coleman's shop were among its core leaders.[125]

The institute educated mechanics and supported their networking and businesses. Its Committee on Instruction organized lecture series on chemistry and mechanics, and offered classes in technical drawing, which Escol attended before moving to Upper Darby. The Committee on Exhibitions sponsored fairs where mechanics marketed new technologies. The Committee on Science and the Arts acted as a professional review board, evaluating inventions and offering cash prizes as an incentive to innovate.[126]

The institute assumed and expanded the Philosophical Society's old role in regional development, partly since the latter institution had become little more than an elite social club.[127] In the 1830s, the institute helped design the city's gasworks, investigated weights and measures for the state legislature, and conducted a series of federally sponsored experiments on steam boiler explosions.[128] It maintained close ties to the University of Pennsylvania, the College of Pharmacy, and the city's first public high school.[129] Students from these schools enjoyed free admission to its lecture series and to the library and reading room in its marble hall on Seventh Street.

Although they remained closely tied to Philadelphia's community of mechanics, industrialists, and institutions, Coleman and his sons, twenty-one-year-old Charles and nineteen-year-old Escol, resigned from the Franklin Institute in 1827. After moving to Cardington, they helped found the much smaller Upper Darby Institute.[130] Sellers & Sons continued to do regular business with Franklin Institute leaders such as Matthias Baldwin, Oliver Evans's successors Rush & Muhlenberg, and Samuel Merrick.[131]

With colleagues in the institute, Coleman brought the English utopian socialist Robert Owen to Philadelphia in 1827. He hosted a party in Owen's honor and helped organize his lecture at the institute. Coleman and some fellow manufacturers were attracted to Owen's model of cooperative manufacturing for its potential to alleviate poverty even as machines began to displace humans in some industries.[132] Their interest in his ideas signaled manufacturers' struggle to manage relations between capital and labor as wages came to replace the more personal ties of apprenticeship. Yet the Sellerses and Philadelphians in general did not develop cooperatives, opting instead to continue the pattern of family firms and partnerships with other producers, which helped sustain their influence over economic development as well as their individual enterprises. This is one reason why they focused most of their investments in education for their children.

Like earlier generations, a fundamental part of the education of Escol's generation took place on the farms and in the mills and factories of their fathers, uncles, and grandfathers. Escol and Charles grew up in the workshops and institutions along High Street, while their younger brother Coleman grew up on the farm and in the shops at Cardington. Coleman remembered, "My father [the elder Coleman] began to teach me the use of a pencil when I was about two years old."[133] He also taught him elementary physics, performing "instructions and experiments for my benefit before I had fairly passed out of the stage of infancy."[134] Although Coleman the father died when his namesake was just seven, his widow Sophonisba "placed the tool-box ahead of the grammar" and continued young Coleman's manual training based on the artistic and scientific education that her own father, Charles Willson Peale, had imparted to her.[135]

In Darby, the Sellerses and their neighbors built their own schools for their children. John, the elder Coleman's cousin and owner of Millbourne mill, acquired land for a schoolhouse as early as 1779. His grandson William, when he turned six in 1830, came to live with him, attending the school and learning to work on the farm and mill with his cousins and indentured servants.[136] Two years later, John, his brother Samuel, their children, and some neighbors in Darby erected a new school for William, his brother John Jr., and their cohorts. The elder Coleman provided the land.

The Sellerses and their neighbors made the school an important center for the moral education of both children and adults. It hosted antislavery meetings and lectures. Prominent abolitionists including John Greenleaf Whittier and James Russell Lowell dined and spent the night at the home of William's father John and their in-law and neighbor Abraham Pennock, as did fugitive slaves.[137]

For the Sellerses and other Darby Quakers, schools under their control afforded opportunities to instill particular knowledge and values in their children and other members of the community. The schools continued in this role even when the legislature passed a statewide public school act in 1836. The Sellerses and their neighbors transferred their schoolhouse to the state system, but John became treasurer and director of the Upper Darby school district.[138]

While they maintained control over local public schooling, the Sellerses and fellow capitalists continued to support a set of private schools that prepared their children to lead industry and institutions. Sophonisba (usually called "Sophy") and Coleman's three sons gained different sorts of formal education, though they all trained to become master mechanics. Charles, their eldest, enrolled in the University of Pennsylvania's medical department before their father died, when he joined Escol full time in the shops.[139] In 1842, Sophy sent fifteen-year-old Coleman to an academy where the lectures covered hydraulics, hydrostatics, pneumatics, optics, galvanizing, electricity, trigonometry, and bookkeeping. With several schoolmates he organized the Franklin Debating Society, taking on capital punishment, intemperance, and other topics.[140] In his free time, he visited nearby workshops and quarries and collected geological specimens. His mother warned he "had better not be buying any more of those cheap books which are now so plenty as they are all trash," referring to the romantic novels of the day. Instead, she sent him a dictionary and encyclopedia of arts and sciences for Christmas.[141]

The educational institutions the Sellerses initiated and frequented in the early nineteenth century furthered their ability to influence economic development. They established another layer of a regional institutional complex designed to foster and regulate industrialization. The new educational landscape enabled increasing specialization and technical sophistication among metropolitan Philadelphia's high-skilled industrial workers and prepared its many working-class communities for careers in manufacturing, transportation, and other industrial labor markets. Public and private education would continue to play a large role in the social organization of rapid urban and industrial growth across the nineteenth and early twentieth century.

The early history of the Sellers family in Pennsylvania illustrates the ways that industry helped drive urban and economic development in the contexts of British colonization and early national America. The dense web of social, institutional, and business networks the Sellerses fashioned with their allies structured the region's internal and external economies. They planned and built the foremost center of industry in the Americas, a status Philadelphia would relinquish only at the end of the nineteenth century.

———————————— ✕ ————————————

CHAPTER 2

MIGRATION STRATEGIES AND INDUSTRIAL FRONTIERS

Around 1820 a German coppersmith named Henri Mogeme arrived at Sellers & Pennock's engine shop on Market Street. Escol's father, Coleman, trying to decide whether to hire him, inquired about his experience. After learning to use tools at the Heidelberg Polytechnic, the visitor replied, he had worked in machine shops in London, Manchester, Liverpool, and then at West Point and the Allaire Iron Works in New York City. Escol remembered that "Father made some remark about his wandering habits," to which Mogeme replied, "Yes, but the more move the more learn and the better work do."[1]

This process of moving, learning, and improving production helped integrate and expand the industrial economy of early and mid-nineteenth-century Europe and North America. It also forged a transatlantic community of skilled mechanics, who like laborers were a highly transient population.[2] The networks and institutions that bound them together and influenced the geography of their movement included business relationships among firms, as well as technical societies such as the Franklin Institute and corresponding institutions in other cities. Family ties helped structure links between firms and regions, too, as elite manufacturers like the Sellerses sent their children to visit and work in factories from England to New England. Since personal contact and direct experience were critical to forming relationships and learning from other places, mechanics' movement itself—whether visits or longer-term migration—was necessary for the diffusion of industry and to sustaining the relationships between firms and cities.

Mechanics approached migration as a strategic enterprise and often served the interests of their cities as well as their families and firms. This was particularly so when their travels aimed at building knowledge, connections, and business for the factories and cities to which they planned to return, even if plans frequently changed. Mechanics of course also shaped the material links between regions. In collaboration with other urban capitalists, they designed, promoted, built, and operated the railroad, telegraph, mining, and factory infrastructure that spread densely across the urbanizing North and more sparsely in the South in the mid-nineteenth century. Both a cause and consequence of these activities, the fortunes and status of East Coast cities became increasingly entwined with their ability to bring regions of the West and South into their orbit.

The travels of the extended Sellers family reveal some of the cleavages between mechanics' aspirations, expectations, and experiences of industrialization in different regions, especially in the antebellum period. As Escol's story suggests, in the South and West their plans rarely translated into the sorts of economic development he and other eastern capitalists envisioned. The Sellerses and other elite manufacturers tended to view industrialization—manufacturing, infrastructure, and public schooling—as the most promising alternative to the slave economy and the poverty of the working classes. Yet their encounters with Catholic immigrant laborers, southern slave owners, and new cities in their travels exposed the limits of this view. Some of these experiences sharpened the Sellerses' views of class and moral economy. At other times they compromised their religious and moral commitments in pursuit of personal and to some extent regional and national interests, to the point that some members of the family used slave labor. Taken together, the family's travels show how industrialization helped produce integrated yet distinct social and material geographies in mid-nineteenth-century America and its systems of cities.

Atlantic Crossings

As a social class, elite mechanics like the Sellerses were a small and relatively cohesive group. In his reminiscences, Escol made it clear that most of the master mechanics in the English-speaking world, numbering several hundred or so by the 1820s, knew one another by name and reputation.[3]

This did not stop them from spying on one another and stealing trade secrets, even from other mechanics in their own family. Not surprisingly, sometimes this inspired conflict. But it was vital for the spread of mechanical knowledge and production capacity, and for firms' and regions' competition with one another.

Henri Mogeme's sojourn at the Sellerses' shop illuminates some of the motivations and tensions shaping this process. By Escol's account, the German "had not a single friend among his brother workmen," who wanted to beat him up and whose suspicion led them to spy on Mogeme. The source of their animosity and mistrust may have had something to do with discord between native and immigrant workers, even at the high-skilled end of the labor market, or it may have reflected class tension. In any event, they learned that he and a boy boarded at a tavern, where they kept drawing instruments, a drafting board, and affixed an "unpickable" lock to their door. That was as much as they learned for certain.[4]

The workers approached the city's mayor, an "intimate and frequent" visitor at the Sellerses' Market Street house, relating that they suspected "Mogeme of being a counterfeiter and forger." This was especially relevant since Sellers & Pennock were presently making machines for the U.S. Mint. They "made out a very clear case that the man's expenditures were far beyond his daily wages," including regular visits to the theater, museum, and restaurants, and purchases of imported wines and fine tracing paper. When Coleman asked Escol to inquire with the German about these accusations, Mogeme listened calmly, did not offer a rebuttal, and soon moved on to explore the anthracite coal region.[5]

Some months thereafter, Escol was in the Market Street store when an elegantly dressed man entered. His "face was clean shaved and under the rim of his hat dark hair curled in ringlets as if just from the curling tongs of the hair dresser. The man spoke before I recognized that it was Henri Mogeme." Escol took him back to see Coleman, and "History was soon told." Mogeme, it turned out, was the son of a duke, and was on his way home since his father had died. He had concealed his identity in his journey through leading workshops of the East for practical motives, he claimed. The only way to gain useful experience in American industry, he believed, was "under an assumed name and character," since "he knew the Americans' fondness of todying [sic] to titled foreigners and . . . society claims would have taken all his time" had he traveled under his real name.[6]

Mogeme carried "a roll of papers in his hand which he unrolled and showed us colored drawings of Hydraulions in detail. He said this was the work of his nephew," the boy with whom he traveled. "In showing his nephew's drawings it was to show how well he had played his part," Escol wrote, "as they had a large number of drawings of American machinery which he said might be useful to him."[7] Revealing their success in copying the Sellerses and their tools was evidently acceptable behavior. At least Escol's reminiscences do not suggest any offense taken.

The Sellerses traveled for their own benefit as well, and often collaborated with other businessmen in planning and paying for their voyages. Like other nineteenth-century Americans from affluent families engaged in everything from architecture to finance, mechanics with means commonly traveled to northern Europe—most often Britain, France, and Germany—to learn and compare current practices while making business connections. In 1832, Coleman and one of his main customers, automated paper mill owner Joshua Gilpin, sent twenty-four-year-old Escol to England on one such trip. His principal mission was to learn all he could about the papermaking, drying, and bleaching machines of London manufacturer Bryan Donkin, though he also took advantage of the broader opportunity to study British industry and technology. Escol took a ship to Liverpool and, by his own account, "spent considerable time previous to going to London in visiting mines, iron and general mechanical works in the vicinity of Manchester, Birmingham, Sheffield and Halifax."[8] In the capital, he met Donkin and toured his factory. He also visited the Maudslay Works, Britain's leading machine builder; Marc Isambard Brunel, who was then building the first tunnel under the River Thames; and his grandfather's old supplier John Mathews, who showed him some new techniques for weaving wire.

Escol wrote regular letters home to Philadelphia with descriptions and drawings of the machinery he encountered. Sellers & Sons' subsequent correspondence with customers regularly referred to ideas Escol had picked up on his trip. Even as he gleaned information from the mechanics with whom he visited, his hosts were equally interested in the American innovations that Escol recounted as they compared shop practices. Learning, however, was not the only mission of the voyage, as the young engineer carried with him a pulp dresser—a machine for removing lumps from pulp—patented by his father, which he marketed to leading British papermakers. Richard Ibotson, whom he visited at his mill on the outskirts of London, purchased the English patent rights to the machine.[9]

There is no evidence that Escol recruited labor during his visit, as many American manufacturers did in this era, but otherwise the trip followed the pattern of Americans in Europe to learn and cultivate business relationships. He stayed in England several months. Soon after his return he married his Philadelphia sweetheart, Rachel Parrish, the daughter of a hardware merchant who lived two blocks from the High Street house where Escol grew up.

Family Networks in the East

The Sellerses' story intersected the other great transatlantic exchange, namely immigration, through marriage as well as in business. Escol's second cousin Mary would marry Edward Bancroft, the son of one of the region's more successful immigrant textile millers, John Bancroft, while Edward's brother Joseph would marry Mary's aunt Sarah Poole. Mechanics in the Sellers, Bancroft, and Poole families would work together in several firms between Delaware and Rhode Island. The three families' migrations illustrate the ways in which mechanics moved and integrated into the workshops and manufacturing sectors of the Mid-Atlantic and New England, as well as some of the social tensions they encountered in the process. It also suggests much about mechanics' motives for settling in particular parts of the industrializing Northeast.

In 1823, John Bancroft left his home in Manchester, Britain's foremost textile center. He settled along the Brandywine Creek on the edge of Wilmington, Delaware, where he set up a flannel mill with his son Samuel. His son Joseph remained in England to finish his apprenticeship in a cotton manufactory, but he joined the rest of the family on the Brandywine the following year, finding work as the superintendent of a nearby cotton mill. The Bancrofts made connections with Quaker millers and manufacturers between Wilmington and Philadelphia, including William Poole and Nathan Sellers and his brother John.[10] Their extended family and allied businesses would structure networks of financing, credit, labor, production, and information in the textile and machine sectors.

Immigration helped grow America's population and networks of skilled mechanics and manufacturers in the nineteenth century, much as it enabled the early milling sector of the colonial era. With prior training in the precise numerical calculation of machine settings for spinning and weaving, the Bancrofts added valuable knowledge and productive capacity to the region's early textile sector. Like papermaking, America's textile industry depended upon the temporary and permanent migration of millers, mechanics, and skilled weavers to build and manage networks of labor and information.[11] Delaware developed a large milling sector to process cotton thanks to these newcomers as well as its easy access to transportation and trading networks radiating from neighboring Philadelphia.

The small scale and precarious financial position of most textile mills in the region made for low barriers to entry, many short-lived firms, and frequent job changes for workers at all levels. By 1827, four years after he arrived in America, John Bancroft abandoned his flannel mill to take charge of a cotton mill owned by Franklin Institute president James Ronaldson along Ridley Creek in Delaware County. Joseph moved with the family but the following year returned to the Brandywine, where he married Sarah Poole.[12]

Working for other manufacturers enabled some skilled mechanics to accumulate sufficient capital to start their own firms. By 1831, the Bancrofts had amassed enough for Samuel to purchase 167 acres with old snuff mills and sawmills in Delaware County, while Joseph bought a mill on the Brandywine. On his property, Samuel erected a three-story woolen mill and boardinghouses for workers. Joseph remodeled his factory to produce cotton textiles, installing some fifteen hundred spindles in addition to looms for weaving.[13] A report by the U.S. secretary of the treasury the following year listed Joseph Bancroft & Co. with $42,000 capital and markets in Philadelphia, Baltimore, and New York City. The firm purchased some of its machinery from Lowell and other Massachusetts shops, but it bought most from the Matteawan Manufacturing Co. on the Hudson River.[14]

The Bancrofts got much more than machines from Matteawan, as Joseph and Samuel's younger brother Edward and their brother-in-law John Morton Poole (called Morton) arranged apprenticeships there in 1830. Matteawan sat on the continent's most important route between East and West: between New York City and the Erie Canal. Like most Hudson Valley towns, it was built by capital from Manhattan. The company's owner, Peter Schenck, had founded the American Institute in Manhattan, an association dominated by industrialists friendly to the administration of John Quincy Adams and its high tariffs on imported manufactures. Poole and Bancroft arrived in time to witness the erection of a new machine shop and foundry that by the 1840s employed close to two hundred men and boasted an annual production of textile machinery, sugar milling machines, and steam engines worth over a quarter million dollars ($6.5 million in 2010).[15]

From Matteawan, Poole kept up a regular correspondence with his sister and her husband, Joseph Bancroft. When the shops took on a job building twenty looms for Joseph's mill, Morton sent home a critical report. One machine was "fitted up in a superior stile [*sic*] for the fair" at the American Institute in New York City. But since Matteawan's bid "was verry low there has not been as much pains taken as ought to have been" on the others.[16] Most of Morton's letters concentrated on news of business conditions and on leaking detailed information about technologies and production practices that might prove useful for Bancroft's business.

Manufacturers of this era worked at stealing more than trade and technical secrets from other firms. In August 1832, Morton wrote Joseph, "Week before last there came here from the neighbourhood of Hartford a man named Kellog[;] he wanted castings for machineary [*sic*] and to kill two birds with one stone he would hire a few hands out of the shop." But when the Matteawan Co.'s superintendent got wind of this plan, he "shew'd him the door and told him to be gone without his castings."[17] Factory owners and managers often worked to prevent as well as to spur migration.

In the same letter, Morton related, "We had the honour of a visit from John Q. Adams," whose host, "instead of bringing him to the office and introducing him," took "him directly to the machine shop."[18] Like the unfruitful voyage of the Connecticut manufacturer Mr. Kellog, the arrival of the ex-president suggested the national importance of the Matteawan Co. Whether for gaining specialized experience, pilfering information or workers, or influencing federal policy, leading firms like Matteawan played key roles in the social and political, as well as the economic and technological, development of industrial America.

Like the German Mogeme, Morton Poole and Edward Bancroft kept moving and learning. In 1833, they left the Hudson Valley for Philadelphia. Edward found work with Coleman Sellers & Sons.[19] Morton attended lectures at the Franklin Institute and took its technical drawing and science classes.[20] In the spring of 1834, he and Edward were back on the road to visit the machine shops of New England, with a loan of $100 and letters of introduction from their families' Brandywine manufacturing colleague, Charles du Pont.

With strong mercantile ties to the cotton-producing South and a proliferation of banks investing in mills, New England's major seaports were developing significant machine-making sectors to serve the nation's first railroad boom and the growth of mill towns along the region's fast-falling rivers.[21] Finding a market for specialized equipment among the banknote engravers, public institutions, and textile mills from Rhode Island to Maine, Poole and Bancroft settled in Providence, a center of machine production at the mouth of the Blackstone River. They remained there nearly three years.[22] Morton was evidently still the more faithful correspondent of the two, writing Joseph Bancroft regularly, seeking advice on textile machinery and other technical problems in the workshop.[23]

Edward and Morton served some of the main drivers of the Providence region's economy. The engraving side of their business provided materials and services to the growing financial sector, mostly in the form of custom banknote plates and dies. By 1830, Rhode Island was already home to forty-seven state-chartered banks—compared to thirty-three and thirty-seven in the much larger states of Pennsylvania and New York, respectively—and most of these were concentrated in Providence.[24] "Taking every thing into consideration I don't see but what we shall be able to earn a comfortable living," Morton wrote Joseph a year after they arrived.[25]

Success in business notwithstanding, Morton struggled with the social changes afoot in Providence. Writing to his sister Sarah, he complained of being stuck in a city buzzing with the religious revival of the Second Great Awakening, "right amongst bigoted, ignorant, zealous fanatics in religion." His contempt for "the sewerish multitude . . . so completely subservient to their ministers," revealed the discomfort and disdain that Quakers like the Pooles and Sellerses felt as they witnessed the social movements of the burgeoning working class.[26] In ascribing this craze solely to the institutions of Yankee New England, Poole failed to recognize that this same revival gripped the mill districts of Delaware and Pennsylvania. He also seemingly did not grasp that mill owners and others of his class encouraged and employed religious revival to engender workers' acceptance of wage labor, regimented factory work, and temperance.[27]

The social schisms of the era sometimes manifested themselves on the shop floor, revealing tensions between natives and immigrants, Protestants and Catholics, and workers of different classes. Henri Mogeme was but one of many immigrant mechanics upon whom colleagues wished violence.[28] At Matteawan, Morton had witnessed an English master mechanic beaten up by the man he replaced.[29] However, as elite mechanics with family firms of their own, the Sellerses and Pooles worked in a highly specialized industry where they were relatively isolated from the lower classes. Their wealth, skills, and consequent mobility meant that they usually traveled in privileged circles.

In 1836, Morton and Edward determined to leave Providence, continuing their tour of "approved establishments" where they could gain new knowledge and skills.[30] Morton secured an offer through his cousin Mary Sellers (who would marry Edward four years later), to become foreman in the Cardington locomotive shop of her cousins Charles and Escol.[31]

The network of Sellers, Bancroft, and Poole mechanics hence tightened further, though Morton's employment in Upper Darby also illustrates some of the less unifying aspects of extended-family capitalism.

Poole viewed his situation at Cardington as another opportunity to learn and spy for the Bancrofts. He approached the Sellers firm much like he treated Matteawan, as a rival from which to glean technical information for the ultimate profit of his own business. Writing to Joseph in February 1837, he warned, "It will be absolutely necessary that Chas & Escol should not know that I have any connection with thy brother in this place for if they did I should never have their confidence."[32] Meanwhile, Edward traveled to Boston, Lowell, and other manufacturing centers in New England, sending detailed accounts of their machine shops to his partner in Upper Darby, before settling back in Rhode Island.[33]

Morton, however, did not return to work with Edward in Providence. In 1838, he started his own shop, making milling machinery in the basement of Joseph's mill on the Brandywine.[34] In his new business, Poole took in his fourteen-year-old nephew William Sellers as an apprentice.

Apprenticeships were an important part of mechanics' extended family networks. When his sister Sarah gave birth to a son in the summer of 1835, Morton sent a congratulatory letter from Providence that, despite its playfulness, described the actual career trajectories of the boys born to the Bancroft, Poole, and Sellers families in the nineteenth century. "I dont intend to say any thing that can be construed into any thing like advice for my dear sister it would be presumptuous in me," he wrote. "However I mean to ask dont you want to make a machine maker of him, we want an apprentice now if you want to bind him out till he is 21." This was no common offer, Morton continued. "Here is two of the verry best masters in the country who are willing to take him and make a man of him[;] had you not better accept the chance you will never get an other so good." Keeping apprenticeship in the family helped ensure decent treatment, if not great comfort. Morton quipped that his newborn nephew "shall share as just we do . . . crackers and cheese eaten on the vice bench and sleep in the same bed that is a mahogany board and the carpenters bench and shall have a share of my old cloak for a coverlid."[35]

On the Brandywine, Poole's apprentices included the sons of several elite Quaker families. William Sellers's shop mates included Barton Hoopes and James Moore, future proprietors of prominent metalworking firms in Philadelphia.[36] Like Morton and Edward at Matteawan, these boys got to participate in setting up the shops when a great flood in January 1839 washed out the basement workshop.[37] The remainder of

William's seven-year apprenticeship was taken up with learning to make and repair textile machinery, including equipment run by steam engines as well as waterpower.

In late 1845 or early 1846, as his younger brother George arrived on the Brandywine to apprentice in Joseph Bancroft's mill, twenty-one-year-old William left one family firm for another. Already a skilled mechanic, he took charge of the machine shop of Fairbanks, Bancroft & Co. in Providence, in which his brother-in-law Edward was a partner. There, he gained experience managing the shop floor of a firm that produced mill gearing as well as stationary and marine engines for factories and ships. He also had occasion to work with a young mechanic named George Corliss, who entered Fairbanks & Bancroft's drafting department and later became a partner when Bancroft joined Sellers in the Philadelphia firm that would become William Sellers & Co. The connections William made in his moves to work would link the leading engineers and manufacturing centers of New England and the Mid-Atlantic in later decades.

This process of skilled migration perpetuated and strengthened the social, technological, and business networks that fostered intensive industrialization in the early and mid-nineteenth century East.[38] Meanwhile, William's second cousins Charles, Escol, and Coleman extended those networks to the West, leaving Philadelphia in the 1840s to develop industrial systems of production, transportation, and cities beyond the Appalachian Mountains. Their travels were born more of economic crisis, and they revealed more nascent and speculative forms of urbanization and industrialization.

Western Systems of Cities

Sellers & Sons was among many firms that did not survive the depression that followed the financial panic of 1837 and lasted into the 1840s. Charles decided to forsake the Cardington factory first, to become a farmer and take advantage of the federal government's offer of land in the West at $1.25 per acre. Escol had little choice but to abandon the firm, and he "joined Charles' western scheme," but with different intentions. Nicholas Biddle, the disgraced president of the defunct Bank of the United States, "owned the controlling interest in the copper mines near Mineral Point, Ill., then idle," Escol later wrote. "He made me inducements to examine and a liberal proposition if I reported favorably."[39]

The opportunity to capture the vast territory and resources beyond the original thirteen colonies inspired manufacturers and financiers at least as much as the farmers and prospectors who dominate popular images of the nineteenth-century West. Also contrary to those images, this was a metropolitan project.[40] Cities, railroads, and their mechanics played central roles in integrating and ordering a national marketplace wherein they turned resources into commodities and brought them to market by coordinating agriculture, mining, processing, and transport. Escol and soon his brothers sought to play a major part in this project of establishing industrial capitalism as the leading system giving order to the economy and cities of the new American territories.

Like many frontier concerns, manufacturing and infrastructure schemes often failed, and this is much of the story of the Sellerses' work outside Philadelphia and the East. However, they and fellow mechanics ultimately transformed the nation's economic geography. They did so by extending networks of skilled labor and production across the continent, diversifying and linking the economies of the cities where they settled. Like their ancestors, Escol and Charles focused on basic products that enabled growth of manufacturing, cities, and metropolitan infrastructure. They would spend the rest of their working lives in the Midwest and South.

Leaving Philadelphia for southern Illinois in the spring of 1841, the two brothers and their wives and children sent their household goods to St. Louis via Pittsburgh, but their journey stalled in Cincinnati, where Charles fell ill. While he recovered, Nicholas Biddle suffered another bank failure that summer, leaving Escol with no copper mines to explore. They decided to stay in Ohio, and had their possessions shipped there.[41] His agricultural scheme on hold, Charles divided his time between work as a physician and mechanic. Escol concentrated on manufacturing.

Though an accident, Cincinnati proved a logical destination for the Sellerses. On the banks of the Ohio River, by 1840 it was the foremost gateway city funneling agricultural surplus from the Ohio Valley to New Orleans and distributing manufactures and commodities from eastern seaports to the growing markets of the West.[42] It was the sixth-largest city in the nation. From a village of 750 souls just four decades earlier, it had become a bustling center of more than 46,000. Local boosters dubbed it the Queen City of the West and projected that by 1890 it would surpass New York City in population.[43]

It was also the largest industrial center in the West. To process produce and raw materials arriving on riverboats, and to serve the city's burgeoning population and trade, entrepreneurs eagerly poured capital into manufacturing. In the federal census of 1840, Cincinnati reported 1,594 manufactories employing 10,608 men and women and producing goods worth some $17 million. In addition to processed pork, which earned it the moniker Porkopolis, the city's manufactures included flour, hemp, furniture, and plows; shoes, hats, saddles, soap, candles, and other by-products of meat processing; and commodities born of southern agriculture such as tobacco, snuff, and cotton textiles. Local metalworkers and machine shops relied on iron ore from western Pennsylvania, Kentucky, and Ohio. Relative isolation beyond the Appalachian Mountains protected these enterprises to some extent from competition with established eastern firms. In 1828, these producers founded the Ohio Mechanics Institute, with a library, lectures, and classes in mathematics, chemistry, geography, and architectural drawing.[44]

Charles and Escol quickly found an investor and partner, local merchant Josiah Lawrence. Their first line of business together was a lead pipe works. By 1844, the three partners opened a larger factory, the Globe Rolling Mill and Wire Works. Their early products included plumbing supplies to bring running water to western cities, telegraph wire for the communication networks linking the East and West, and iron rails for railroads—in short, basic materials for an industrializing region.

The Sellers brothers aspired to reach national markets from Cincinnati. Escol traveled to New York in 1844 in search of credit, customers, and suppliers. Charles negotiated with plumbers from Pittsburgh to serve as agents selling their pipe in western Pennsylvania. He wrote to Escol, "By securing them we think we have secured the Pittsburgh market. We have also had an application from a plumber in Stubenville for the agency" in markets up the Ohio River.[45] The machine they built to make lead pipe worked well, Charles wrote their sister Anna. "We have made as much as 600 [pounds] in an hour. . . . It is exciting a great deal of attention all over the country, applications are being made to us continually for the agencies for selling our pipe in different places, even as far off as Baltimore." Beyond the East and West, he added, "we are in hopes of being able to secure the monopoly of the New Orleans market, Mobile and perhaps Havana."[46]

Cincinnati may have been a strategic location, but it was not an easy place to run a large, up-to-date mill of the sort the Sellerses had come to expect back East. One key obstacle was "not having capital to buy sufficient stock of lead to push the business as much as we would like," Charles complained.[47] The Sellerses were accustomed to Philadelphia's easy access to bankers and merchant capital, wholesalers procuring supplies, and specialized machine makers with whom to form networks of production and share its costs. Charles wrote their mother, Sophy, in September 1845, "I was unexpectedly obliged to make a tour among the Iron works on the Cumberland River to make arrangements for a supply of Blooms for our Roling Mill."[48] This was his first trip south on business, and it opened his eyes to the differences between regions.

Charles's main complaints concerned the urban and industrial infrastructure of Tennessee (he said nothing of slavery, nor would he in subsequent letters). "Nashville is a very pretty city but I should not like to live there," he told his mother, since it "had not a comfortable feel to me." Much worse, "The Iron region of Tennessee is about as rough a section of country as I ever was in." Unlike the mining and mill towns of Pennsylvania, which were linked to Philadelphia by well-worn roads and increasingly rails, "I had to travel from furnace to furnace on horseback through dense forests without roads. Nothing but a narrow path like a foot path."[49]

Back in Cincinnati, Charles also found his work stymied by the lack of reliable metropolitan infrastructure and business networks necessary for the sort of advanced production he and Escol were trying to establish. Supply lines dependent on water transport threatened to stall production, as "the river is so low that we are afraid we will not be able to get a stock of Iron and coal before what we have is all worked up."[50] Two months later, in November, iron shipments were arriving, but several pieces of heavy equipment had broken, including two furnaces and the shears that cut the metal. Charles made these himself, since Ohio lacked equally skilled metalworkers who could readily supply parts. When he returned to Tennessee the following spring, he brought his youngest brother out to Cincinnati to substitute for him at the mill. Coleman reported similar frustrations to their mother. "We are now very much at a loss for water," as the city's "water-works gave out some weeks ago," he wrote, and "all the water for the engine has to be hauled from the river in barrels."[51]

Their financial backing remained shaky as well, a symptom of larger structural challenges manufacturers faced in the West. In 1845, in a letter to Escol who was again in New York, Charles revealed that their partner was having second thoughts. "Last week Mr. L had an awful attack of the Blues. I think mainly attributable to a bad cold. He wanted me to write to you to see if you could not get some [funds] out East to buy him out." Compounding this uncertainty, a shortage of specie in the chronically cash-poor West made it difficult to pay workers and suppliers. "Money is in great demand here by the produce dealers and none to be had even for the best and shortest paper" promissory notes that could be easily redeemed in an economy that mainly ran on IOUs. "The banks are entirely run out."[52]

The good news was the mill had just received an order from the Madison Railroad for two thousand tons of iron rails, and the large capital that investors poured into railroads would make their mill a going concern. "I have no fears of being able to sell all the Iron we can make at the highest prices," Charles declared.[53] As one of the most advanced metal manufacturers of the West at the time, they attracted an expanding flow of orders from midwestern railroads and telegraph lines.

This surge in business inspired Charles and Escol to send for Coleman on a frequent basis, and in 1849 he settled in the city. Their cousin John Sellers Jr. also arrived from the mill districts of Delaware County. Like Charles he worked in two sectors, at a grain mill as well as the Lafayette Bank.[54] In early 1850 John Jr. would return to Philadelphia and three years later take charge of financial management at his brother's machine tool works, William Sellers & Co.

A steady stream of letters plus frequent visits, especially of younger cousins and older aunts, afforded regular opportunities for the Sellerses to compare Cincinnati to Philadelphia. On a Saturday in July 1849, Coleman recorded in his diary, "We had a great argument this evening about the relative advantages of different cities as a place of residence." In their debate, he and his mother Sophy, who was visiting, took "the part of Cincinnati," with his aunt Anna and cousin John "arguing in favor of Philadelphia. John thought that the mosquitoes were an objection to this place. We told him that Cincinnati was far superior to Philadelphia in even that respect. We had the largest mosquitoes and they bite the hardest."[55]

During her first stay in the summer of 1846, Sophy wrote a detailed account of Cincinnati's environment and social life to her sister-in-law Hannah Hill back in Delaware County. She found the city's homes and gardens quaint, though they also reflected its rapid industrialization. The granite of their front steps "gives the city a more somber look than our white marble [in Philadelphia] in consequence of the smoke blackening every thing so much it gives them quite an ancient look[;] the turrets and steeples of the churches are all blacken'd." Although the city lacked the street paving and water systems of older places, on the positive side goods like food and clothing were often, in Sophy's view, as good as back east. "I have been in some stores quite equal to some of our best stores," she professed.[56]

Sophy's reactions to the habits of women and children in Cincinnati suggest her expectations of middle-class behavior went unmet. "Escol has a snug 2 story house . . . quite out of the fashionable part of the city," she wrote, in a neighborhood where some behavior seemed "the same as in the country." For instance, "Children of the first families run about barefoot."[57] Local women covered the wrong parts of their bodies. "In walking out after sunsett scarce any one wear bonnets sometimes they carry a sun bonnet in their hand every one makes themselves as comfortable as they can." However, they went "into the extreme of wearing a great many skirts and their dresses so long that they sweep the streets thus doing away the necessity of sweeping them by the scavengers who once in a while scrape the streets."[58]

Beyond mosquitoes, the animal life of cities in the nineteenth century presented one of the greatest challenges to middle-class families, forcing them to confront problems of public health and personal comfort. The scavengers were the pigs that roamed the city's streets, ensuring that "the moment any slop is thrown out it is cleaned away directly. This we cannot like," Sophy declared, "as the consequence resulting from so many pigs fleas are abundant & our Anna [Escol's daughter] suffers severely from them as their bite raises large welts on her." Charles's family rented a house in a healthier situation, "on one of the hills which bound the city to the north in a snug little cottage." They had a "large garden in which they raise their vegetables [and] a number of fruit trees," and "they have very kind neighbours who are ever ready to supply them with any thing they want."[59]

Of most concern to Sophy was the toll that work took on her sons, as they toiled longer hours at heavier work than they had in Philadelphia, where they had been part of a much larger pool of skilled labor. Charles, Escol, and Coleman, she wrote, "go to the mill a little after 6 and seldom get home before quite dark—often 10 or 12 at night." Though her youngest "likes the business and Cincinnati," she found him "very thin" and with his "dirty business his clothes are in such a state [in] this warm weather that it is scarce possible to make them clean[;] to keep him any way decent keeps me pretty busy." Like his brothers, Coleman was absorbed by work: "he seems so wrapped up in it that he has not formed any acquaintances[;] when he comes home after eating his supper he cares for little else than a book."[60]

Coleman and his brothers did explore and learn more in Ohio than Sophy's worry suggests. Fascinated by the natural landscape of Ohio, Coleman spent his occasional days off studying the topography and flora of the countryside. "You should be astonished to see the emense variety of wild flowers the woods . . . contain," he wrote his mother.[61] He

accumulated rocks, shells, and other specimens of natural history, prompt-
ing one relative to observe that his "room is itself a museum."[62] Escol ex-
cavated Indian burial mounds in the region and collected ancient artifacts.
Throughout their lives, the brothers gleaned what nineteenth-century
Americans called "useful knowledge" from their travels, work, and virtually any activity in which they engaged. This was particularly so with
Coleman. Escol harbored somewhat greater business ambitions.

By the late 1840s, Cincinnati had a substantial community of skilled
mechanics, including many like the Sellerses who had migrated from
the Northeast and maintained ties to national networks of mechanics.
All three brothers participated in the local Mechanics Institute, of which
Coleman wrote fondly in his diary.[63] They regularly received and pored
over the reports of the Smithsonian and other technical societies of the
East. Family members visiting and writing from Philadelphia, Delaware,
and Providence related news of the social and mechanical happenings
across the Appalachians.

However, mechanics were a small minority in a diverse city dominated
by newcomers. Cincinnati's population was 46 percent foreign-born in
1840, mostly recent immigrants from Germany and Ireland. The census that year counted twelve hundred Pennsylvania-born men, more than
those from Ohio. Many came from New Jersey, New York, Virginia, and
Maryland, a mishmash of industrial North and slave South. Blacks made up
5 percent of the population, about as much as whites from Pennsylvania.[64]

Cincinnati was a city where people with competing values and visions of economic development came into regular and intimate contact.
Historians have characterized its capitalists as both "exuberantly committed to exemplifying the potential of free men working in a free society
and shamelessly dedicated to placating its good customers of the South."[65]
Their warehouses sat just across the river from Kentucky, and the city's
growth owed much to customers, suppliers, and slave labor in the South.
This was true of manufacturing as well as commerce, including Escol and
Charles's business, and in the late 1840s they would work more directly
with slaves in southern industry.

In the city, the social tensions and contradictions the Sellers brothers
experienced, and often embodied themselves, reflected the competition
between different groups' visions of class, economic development, and
the state. Like Morton Poole in New England, they looked with suspicion
upon recent immigrants who worked under them in the factory and in
other lower-paid and irregular work in warehousing, shipping, and domestic service. These people threatened the stability and what Coleman

viewed as a moral economy of industrial capitalism and the institutions that supported it. Writing in his diary in 1853, in the heyday of American nativism and know-nothingism, he captured some of the converging tensions between labor and capital, Irish Catholic immigrants and Protestant natives. "There has been strikes among all classes of workmen for higher wages and a general excitement through town about the comming election," he wrote. "The great opposition of the Catholics to our system of free schools has made this a point on Politics, and the strife seems high between the Pope's emmisarys and American freemen. God grant that our noble free schools may not be destroyed."[66]

By the 1850s, for elite eastern industrialists like the Sellerses, a free urban society generally meant one in which public schools prepared a transient, often immigrant population with a standard set of reading, writing, and social skills to work in industry.[67] By "all classes of workmen," Coleman did not mean mechanics like himself, who supervised entire factories or departments within them and typically arranged an annual salary with the firm's owner. Rather, the people striking were mainly immigrants and Catholics, whereas the people voting (and most of those promoting public schooling) were native-born white property owners.

Social divisions and contradictions persisted as the Sellerses, like other eastern migrants with means, limited their interaction with other classes. They removed their children from the public system and sent them to private academies and tutors. Charles wrote his mother that his son Morris "appears to take much more pleasure in his studies than while he was going to the District Schools."[68] Coleman's diary during his time in Cincinnati evolved from ambivalent to disdainful of the city's working class, though he made some distinctions between immigrants. By 1854, he would write, "I sent an advertisement to be inserted in the German Paper for a German Girl (Protestant) for our house[keeping]. I do not want any more Irish or Catholic girls they are too deceitful."[69]

These tensions notwithstanding, and perhaps in some ways because of them, the Sellers brothers remained alert to opportunities to shape the development of other regions beyond Cincinnati. They watched as the conclusion of the Mexican War and the acquisition by treaty of the disputed Oregon territory expanded the nation's boundaries to the Pacific. Building the infrastructure to industrialize these new territories presented Escol and Charles with a chance to help define the nation's economic geography on a larger scale, putting places on the map with rail links and large manufacturing plants. "In this widely extended country there are vast tracts of arable land, as well as abounding in mineral wealth," Escol wrote, "which need but the facilities of communication with the great marts in order to develope their value."[70]

Beginning around 1845 and continuing for almost a decade, Escol focused his greatest energies on designing and promoting a new system for railroads to cross not only the American continent but also Europe. His locomotive and the rails on which it would run were lighter than those in use at the time, promising rapid construction and higher speeds of travel. With Coleman increasingly in Cincinnati and Charles working on and off in the South, Escol spent much of his time in New York, since his system required substantial capital and was tied to the progress of massive infrastructure projects that eastern bankers and politicians were in the best position to pull off.

Escol patented the improvements to his locomotive in Britain and the United States in 1847 and subsequently secured French and Belgian patents. He formed an association with Miles Greenwood and two other Cincinnati investors to finance the invention's promotion, sharing the patent rights with these partners. With Coleman and a local mechanic named John Whetstone, Escol built a model of his light locomotive in the Globe Mill. They put it through its first public trial in the city on May 28, 1848. Coleman and Whetstone began studying French, in preparation to exhibit the model in Europe, though their trip was canceled when revolution broke out in France that year.[71]

In late June, Escol and his two young assistants brought the model to New York to demonstrate and market his system. He arranged for a testing track to be laid on a vacant lot at Fourth Avenue between Twenty-Second and Twenty-Third Streets. Beginning in August, they ran the locomotive, pulling various weights, for an audience that included some of the leading railroad engineers of the East. The American Institute awarded the invention a gold medal at its fair that fall.[72]

The system's most important advocates in the East were Horatio Allen of New York's leading machine firm, the Novelty Works, and Philadelphian William Norris. Both men were associated with capitalists seeking to build a railroad across the rugged Isthmus of Panama. Escol spent much of the next two years negotiating to build his system there. In the meantime, he sought out railroad officials in New York, Philadelphia, Baltimore, and Washington in the fall of 1848 and drafted a promotional account of his improvements, securing testimonials from leading engineers and manufacturers.[73]

Escol devoted so much time to this project, and Charles was so consumed with work in the South, that they neglected to sustain the Globe

Works and its core lines of business. They liquidated their partnership with Josiah Lawrence at the end of the summer of 1849. Cincinnati remained their home base, where their families resided, and Escol and Rachel adopted his orphaned cousin Louisa Peale that same year, despite their current financial problems.[74] With customers lacking the capital to pay for their machines, they covered their debts by leasing the mill and wire manufactory to another businessman "with all the priviledges of use of tools & machinery and all the contracts [for] the term of one year at the rate of Three Thousand Dollars."[75] This gave Escol and Charles some capital to tide them over as they pursued railroad-building ventures.

In the national economy of 1849, the timing of Escol's invention appeared fortuitous. The Gold Rush in California brought a heightened sense of urgency for a transcontinental railroad to bring the United States' most important source of specie to the nation's economic centers on the Atlantic coast. He quickly adapted his scheme, drafting a "plan for establishing a rail-road to San Francisco," which he peddled to capitalists in New York and Philadelphia and to politicians in Washington.[76]

Escol's plan called for a "Pioneer Rail-road" of wood rails, using his light grade-climbing locomotives, to be built within five years, and "a first class iron railroad" to be completed within twenty years. He projected eight hundred passengers per day at $100 each, $50 million worth of gold per year charged 1 percent of its value for transit, and four hundred tons of other freight per day at $40 per ton, for total annual revenues of $29.3 million. This, Escol deduced, would generate profits of approximately $14.65 million ($431 million in 2010), of which he assigned $8 million as an annual appropriation for the iron road and $6.65 million for stockholder dividends.[77]

Escol proposed a public-private partnership in what amounted to a national plan. Heralding the public purpose of such a railroad, he outlined a compact to which the federal government and a joint-stock company would each contribute. Congress would provide a "grant of land three miles wide on each side of the road in entire strip," "Protection of the work in its progress and after its completion against hostile Indians and others," and $30 million backed by an issue of U.S. bonds. In return, the company would immediately line the route with telegraphs and finish the wood railroad in five years and the iron one in twenty years. It would carry the U.S. mail and federal officials on business free of charge.[78]

The transcontinental railroad would be completed in twenty-three years, roughly the time Escol proposed. But it would not be his project.

Instead, in 1850, an old classmate from Upper Darby who was chief engineer of the Panama Railroad offered him $10,000 to use his patent and hired him as its mechanical engineer to oversee locomotive and railcar production.[79]

Crossing Panama promised eastern capitalists a chance to dominate the markets of the West Coast and establish themselves in Latin America and the Pacific, but for Escol the deal proved disastrous. When he put out bids to manufacturers to build his locomotives, few seemed interested, and the bids that did come in gave prohibitively high prices. Fearing that his system would not be built, he quit his position and contracted with the company to build three locomotives with Coleman and John Whetstone in Cincinnati, apparently at cost.[80] As they began production, unbeknownst to Escol his classmate resigned from the railroad, and its new engineers found a different route that eliminated the need for his grade-climbing system. When the first locomotive arrived in New York, the company's officers insisted on a trial and refused payment. After much wrangling, the railroad's directors instructed him to complete the other two locomotives but to leave off the machinery required for scaling steep inclines. In the end, they would not pay Escol for his patent. This resulted in a costly court battle won by the railroad.[81]

At the same time, Escol and Charles became embroiled in mostly failed efforts at industrializing the South and at building the light railroad. If their time in Cincinnati showed how challenging it was to establish cutting-edge industry in young cities of the West, their experiences in the late 1840s and early '50s illustrate some of the deepest moral contradictions that northern industrialists proved willing to accept in this era. Both brothers seemed increasingly desperate to succeed in some substantial venture, and especially in Escol's case to realize his own designs for big infrastructure. These are the parts of their lives that most resemble Twain and Warner's protagonist in *The Gilded Age*.

Southern Failures

Charles's and Escol's forays in the antebellum South demonstrate how far they strayed from the abolitionism of their grandfather Nathan and other family members. Their drive to complete large, profitable projects made them willing to participate in the slave economy, a not uncommon pattern among northerners. In the process, the Sellerses took part in the some of the most ambitious and speculative projects of southern industrialization. These experiences reflect some of the most exploitative and vain attempts at regional economic development in the era. They also reveal limits to urban and industrial growth distinct from those in the West.

After leaving the Globe Works, Charles tried to get an iron furnace off the ground in the Tennessee Valley. He and his partners made little progress and accumulated significant debts. They used slave labor rented by one of the men. In his letters to family, Charles apparently did not share his feelings about being a slave master. However, a letter home from Sophonisba during her visit in 1848 reveals how at least one family member sought to rationalize this.

Sophy attempted to cast his actions, if not his position, in a benevolent, paternalistic light. For slaves under his supervision, he replaced violent punishment with incentives for obedience at work and calm in the camp where they lived. "Charles as a stimulus for good behavior allows those that wish it a pint of roasted coffee a week, he having abolished whipping, should they misbehave they are deprived of their coffee."[82]

Following earlier assessments of Cincinnati residents, Sophy used clothing as an indicator of social patterns, writing relatives in Philadelphia, "Charles has pockets put in their pantaloons which has pleased them very much this being a luxury they never had before." This implied that he aspired to a level of trust above that of most masters, who feared that pockets enabled slaves to steal or to conceal weapons to turn on their oppressors. Charles's slaves, his mother concluded, "appear to be very happy, and say they love him more than any master they ever had. Still to me it is a sad feeling to think they are in bondage."[83]

Other family members were unwilling to make or excuse these sorts of compromises, and some were active in antislavery and allied social movements. That same year, Charles's father's cousin John Sellers, William's father, served as a delegate to the Free-Soil Convention in Buffalo. He hosted abolitionists passing through Delaware County on speaking tours and hid runaway slaves in his Upper Darby basement.[84] His wife Elizabeth (Poole) quit the Darby meeting three years earlier, in 1845, over its failure to follow through with "most active and untiring exertions" on its members' "lofty position" against "evils so fraught with misery and degradation to millions of your fellow creatures," namely slavery, war, and intemperance.[85]

Beginning in the 1820s, their in-law Abraham Pennock figured prominently in the region's Underground Railroad network and its woman suffrage, temperance, and free-produce movements. This last cause promoted an antebellum version of fair trade, partly by buying cotton from freedmen and white farmers who did not use slave labor and processing it in a mill underwritten by the Free Produce Association of Friends in the 1840s. However, the movement's public statements about what to do about slavery were moderate and almost naïve in their goal to establish a model of commerce and industry that would induce southern plantation owners to abandon slavery.[86]

Like Pennock and Nathan Sellers before him, John moderated his abolitionism to some extent. According to a friend, John "deprecated the spirit of disunion found in some of the ultra abolitionists, and often said 'the best way to abolish slavery would be to introduce the public-school system in the South.'"[87] Ultimately, most members of the family believed that the institutions and material transformations of the industrial economy they were building held the potential to solve America's social problems. The perplexity, alarm, and disdain with which the Sellerses viewed people of other social classes, like Sophy's partial apology for her son's use of slaves, revealed how they found this stance undermined.

It also pointed up some of the thornier contradictions in their actions and alliances as capitalists. Most damningly, the use of slaves in itself violated any commitment to a moral economy they may have professed to share. Charles's and Escol's institutional affiliations (mainly in technical societies, as opposed to social reform organizations) also reflected this departure from the values of other family members. Charles's particular management strategies in Tennessee were consistent with other family members' approach to manipulating labor at work and at home. But of course southern capitalists would not permit the slaves who toiled in their mines and factories and on their railroads to become an educated, free workforce that could support the sorts of innovation and production that sustained Philadelphia or even Cincinnati.

There was probably no venture in which a greater incongruity existed between visions and realities of southern industrialization than Charles and Escol's association with Duff Green. Escol met him in New York and Washington sometime in the late 1840s, and Green soon tried to lure the Sellers brothers to work for him. In what was his typical fashion, he held out the promise of great engineering works to transform the South. "I have a contract for twelve thousand tons of iron, fifty burden cars, & five locomotives and intend to manufacture the iron cars & locomotives myself if I can get a suitable person to take charge of an establishment," he wrote after Charles's forge failed. In addition to having Charles supervise this contract, Green proposed that Charles help him purchase and improve iron furnaces in East Tennessee.[88]

Duff Green was among the most passionate and best-connected promoters of industrialization in the antebellum South. A newspaper owner, editor, and politician from Kentucky, he married the daughter of the governor of Illinois and gained the title of "General" when President Monroe appointed him a brigadier general in the Missouri state militia. He was a member of Andrew Jackson's "Kitchen Cabinet" before changing sides to support John Calhoun, who had married Green's daughter. As a delegate to the Missouri Constitutional Convention, government mail contractor, President Tyler's special agent in England, France, and Mexico and his consul in Texas in the early 1840s, he helped cut the deals for America's economic expansion.[89]

Green's stationery in the late 1840s advertised his work as an attorney with an outfit called the General Agency in Washington: "Practice in the Courts of the District of Columbia and Supreme Court of the United States. Have made arrangements for the prompt execution of any law or agency business in the Southern and Western States, and the Territories." He pushed his and his clients' interests in these regions' development to nearly every federal institution. He would "attend promptly to Claims before Congress, or any of the Departments at Washington."[90]

Most important for the Sellerses, Green sought to develop the South's natural resources, manufacturing, and infrastructure on a grand scale. His plans included a canal from the Mississippi to the Rio Grande and consolidating all railroads below the Mason–Dixon Line. To finance these designs, he organized the Pennsylvania Fiscal Agency, which northerners reorganized during the Civil War as Crédit Mobilier of America.

For Green and his investors, mechanics from the North promised to overcome the limited supply of skilled factory and railroad labor in the South. Some northern firms, including Philadelphia's Baldwin Locomotive, loaned workers to southern concerns. Others like the Sellers brothers contracted individually.[91] Charles went to work for Green's East Tennessee & Georgia Railroad, overseeing construction of locomotives and freight cars in its workshops at Athens, Tennessee, and Dalton, Georgia.[92] He almost surely used a mix of free and slave labor there, though his letters offer few details.

Green seduced Escol with assurances that tickled his greatest aspirations, and tried to divert his railroad building energies from the West to the South. In February 1849 the general wrote, "Nothing will be done as to the California Rail Road during the present session" of Congress, but "there will be no lack of work as I can control several very large contracts" for railroads in the South. These included at least one "on which I could . . . introduce your improved locomotive."[93] The system attracted Green for its low costs and rapid implementation.

In October he made a bigger offer, telling Escol "I wish to be authorized to dispose of your patent rights for the road & locomotive in Illinois Kentucky & all the states & territories south of Mason & Dixon's line & West of the Mississippi." He claimed, "If you will agree to give me the agency . . . I can immediately make an arrangement for the construction of a Road" in Alabama connecting the Coosa and Tennessee Rivers. He concluded, "If we are [correct] as to the value of your improvement that road of 30 miles will induce other companies to adopt it."[94] In March 1850, Escol headed south from Cincinnati to contract with the East Tennessee & Georgia Railroad, presumably the first step in this larger plan.[95]

But the Sellers brothers were already experiencing the shortcomings of doing business with Duff Green. Charles and Escol's biggest complaint, to judge from their letters, was the lack of cash and credit to keep their projects running. This was an indication of both the underdeveloped financial sector of the South and the overextended nature of Green's enterprises. In December 1849, Charles reported that Green had authorized him to draw on his account with a Washington merchant bank, but that the general's business agent in New York had already withdrawn all the funds.[96] When he proved unable to deliver salary payments for Charles and his workers, Green allowed him to run his sawmill, taking two-thirds of the profits until they recouped this debt.[97] The contractor responsible for operating the railroad was slow in making payments, so Green's payments to Escol for his patent rights also stalled.[98] The following summer, Charles's daughter wrote Escol, he "will not settle fathers accounts he seems determined if he can to cheat him out of it. . . . The people here dislike him very much—many think him deranged . . . there is scarcely a man here he does not owe something."[99]

Like the Sellerses' plans for manufacturing in Cincinnati and for railroads across the continent, Duff Green's grandiose quest to transform the South met with a host of obstacles. His overblown expectations and charlatanry compounded his limited access to cash, credit, and skilled labor. Like Twain and Warner's character Colonel Sellers, his dependence upon appropriations from Congress that were not forthcoming also left the general and his northern collaborators short of their goals. In the broadest terms, however, the main barrier to industrialization in the South lay in the structure of its economy, which was dominated by plantation agriculture and slave labor.

The South also lacked the networks of mechanics and of cities that spurred industrial economic development in the North. Nor, ironically, did it have enough Duff Greens, influential capitalists forming networks of investment, labor, and production, to make this form of development viable. As implied in Charles's remarks about Nashville and its hinterland, southern cities were ill equipped to support industrialization, and not coincidentally they grew far more slowly than the cities of the North and West. Industrial development proceeded most rapidly and successfully where it could depend not only on free labor motivated to innovate and accumulate capital, but also upon diverse markets and strong networks of firms and institutions. These dynamics resulted in uneven patterns of urban and economic development that intensified the differences between North and South.

Outside of more mature industrial cities such as Philadelphia, mechanics like Escol and Charles also lacked the influence to determine the course of development. Their forays in building grand railroads and marketing their own costly inventions were promising but highly speculative. Ultimately, they lacked sufficiently strong social, business, and institutional networks to realize these more ambitious projects.

The Panama Railroad fiasco left Escol dejected and without capital to continue marketing his invention. He gave Whetstone tools in lieu of back wages. In January 1852, Coleman and Whetstone went to work for the Cincinnati steam engine, sugar mill, and locomotive maker Niles & Co. The year before, Coleman had married a girl from Cincinnati, Cornelia Wells, and their first child was on the way, so he presumably needed a steadier job.

To settle his debts, Escol became Niles & Co.'s principal draftsman, sometimes engineer, and traveling salesman. He received some satisfaction when the Coal Run Improvement Railroad installed his rail system on its small spur serving mines in Pennsylvania, with the light locomotives built at Niles.[100] Coleman would go on to much greater prominence as an engineer in the late nineteenth century, while the later careers of Escol and Charles would take more modest turns.

The Maturing Midwest

In their early years in Cincinnati and their forays south, the Sellers brothers were in the business of starting something. By midcentury, they were serving the next phase of industrial development. With Chicago and St. Louis leading the race to be the main gateway for trade between East and West, Cincinnati capitalists struggled to carve out a significant place in the nation's economic geography. Local machine makers became more specialized, mainly to serve the railroads that promised to keep the city relevant. The Sellers brothers' work in the 1850s reflects this shift in focus, though they would all soon move on from Ohio.

Jonathan and James Niles hired Coleman as an accountant, timekeeper, and assistant foreman. When the foreman of the locomotive shop left in February 1853, they made him both bookkeeper and acting foreman. Escol and Charles meanwhile received regular offers for their younger brother to work at railroad shops around Ohio, which helped them negotiate a better deal in which Coleman shed his accounting duties and became foreman at an annual salary of $1,200.

Coleman built up the locomotive shop's capacity, turning out engines for the Indianapolis & Cincinnati, Memphis & Charleston, Bellefontaine & Indiana, and the New Orleans, Opelousas & Great Western Railroads, among others.[101] Even with all this activity, Coleman found time to work on his own inventions, including "a machine to register the running time of all trains their speed [at] each second and the time of all stoppages and startings."[102] This sort of tool could help railroads orchestrate the movement of goods and people between cities and regions in more predictable, systematic ways.

Despite being "hampered continually for the want of machine tools powerful enough to do the work as [he] desired it done," by Coleman's own later account, his superintendence brought Niles & Co.'s locomotive shop up to eastern standards.[103] In November 1854, the eastern trade journal *Railroad Advocate*, reporting on its tour of western shops, gave the firm a rave review. "The West is fast becoming competent for the supply of the equipments of her own roads," and Niles was first among machine makers, it claimed. "On a visit to Cincinnati, we were not prepared to find so large and so well arranged and well equipped works as theirs, engaged in the business. We owe an apology to the West for such a confession, but we had, until then, supposed that nearly all Western roads were dependent on the East for Locomotives."[104]

This praise notwithstanding, Niles & Co. decided to reduce its locomotive production that same year. Coleman noted in his journal, "Mr. Niles says [he] has made up his mind to cut down the force on Locomotive work and only turn out an engine every two weeks. I am very sorry for it, for I have just got the shop organized to turn out one a week with care."[105] He concluded in a letter to his mother the following year, "Locomotive building seems to be a poor business in the West—our market is limited to the Western roads." The small regional railroads Niles served "are all very poor and do not at present seem inclined to find cash to pay for engines when they can go east and buy cheaper machines."[106] The firm, for example, had some $50,000 to $100,000 tied up in Ohio & Mississippi Railroad stock, having furnished most of that line's rolling stock.[107]

Coleman recognized that Cincinnati was becoming a regional economic center, not the national center that his brothers and fellow entrepreneurs imagined in the 1840s. In early 1856, Coleman would abandon the West to join his cousins at the William Sellers & Co. machine tool works in Philadelphia. Although Cincinnati would become a major producer of machine tools later in the century, it declined as a locomotive building center after the Panic of 1857 and put out hardly any railroad engines after the Civil War.[108]

Railroad development in the West, however, proceeded at a frenzied pace, in successive boom and bust cycles. Eastern and western capitalists continued to invest in transportation infrastructure to favor their particular cities in the race for the continent's trade. From its initial railroad boom in the 1840s through the railroad wars among New York financiers in the 1880s, Ohio was the most hotly contested territory in which eastern cities competed to build the dominant link to the West.

In 1853, Charles took a job as superintendent and master machinist of one such gateway railroad, the Bellefontaine & Indiana. The company's annual report that year touted its connections to all the major markets, including "a direct connection with the city of New York, the great commercial emporium of this continent, and with New England, the great produce market of this country." Through its junction with the Ohio & Pennsylvania Railroad at Galion, Ohio, slated for completion that spring, the Bellefontaine & Indiana would "have a continuous Rail Road communication by that route, with Pittsburgh," Philadelphia, and Baltimore. "Thus the two great lines of Road from Philadelphia and Baltimore on the one hand, and New York and Boston on the other, will converge together at Galion," proclaimed its president, "and thence pass united over our Road, and its Western connections through Indianapolis and Terre Haute, to St. Louis, the great commercial centre of the Mississippi Valley."[109]

Charles hastened the production of rails and equipment to ensure that the company's advantageous position would not be undermined by faster development of other lines. Eschewing the inferior iron of Tennessee, he

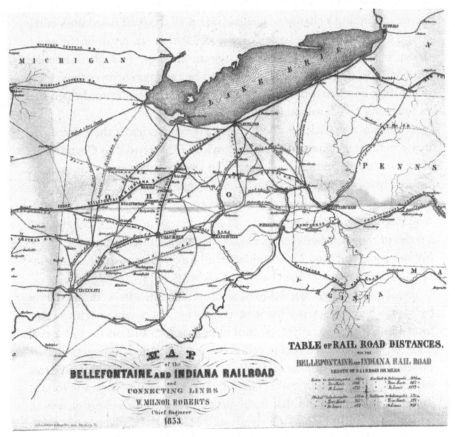

Figure 6. Detail of map of Bellefontaine & Indiana Railroad and connecting lines, 1853. From: *Third Annual Report of the President and Directors of the Bellefontaine & Indiana Railroad* (New York: Leavitt and Allen, 1853). Library Company of Philadelphia.

purchased rails from Britain and spikes from the Albany Iron Works in Troy, New York. He ordered 20 locomotives, 100 gravel cars, 105 horse cars, 40 platform cars, 6 postal and baggage cars, 10 passenger cars, and 45 stock cars, suggesting the large volume and range of anticipated business.[110]

Escol played a major part in this work, seeking to help the railroad solve one of the most important technological problems facing the integration of national transportation systems. The Bellefontaine & Indiana's president explained in the annual report, "At the Indiana line . . . there is a change of guage [*sic*], from 4 ft. 10 inches, to 4 ft. 8 1/2 inches. We are now having a Baggage Car constructed at Cincinnati, under the direction of G. Escol Sellers, Esq., arranged on a plan of his invention, to run on *both guages* [*sic*]." This innovation promised to "obviate the inconveniences attending a change of guages [*sic*] on all the Rail Road lines of the country."[111] Linking the nation's railroads, which ran on different width tracks, meant integrating America's system of cities. Though Galion did not come close to winning the contest to be the great urban gateway between East and West, the railroad and its backers' boosterism helped fuel its financing and development.

Coleman and Charles would spend the remainder of their careers in major industrial centers of the North, while Escol retired to Chattanooga after more disappointment in business. In 1859, Charles wrote his sister, he "settled in a civilized land again," in Chicago, which became the leading city of the West.[112] He lived there for the rest of his life, working as a railroad engineer and photographer, and passed away in 1898. Escol endured a series of tragedies between 1852 and 1860, as his three daughters, one of his two sons, and then his wife Rachel all died.[113] His career also continued to resemble that of the character Colonel Sellers.

Escol's last substantial business venture, in the 1860s, brought him back to where he started his western journey, in a sense. With capital from Philadelphia investors, he moved to southern Illinois to start a coal mine on the Ohio River. He called the place Sellers Landing and envisioned it as a major stop for steamboats to buy coal. Escol and his investors expected urban growth and economic diversification to logically follow. To his dismay, he found that many boats still ran their boilers on wood, which was plentiful along the riverbanks. His ambitions were once again foiled by a slower pace and less ubiquitous spread of industrialization than he expected.[114]

With his coal-vending scheme floundering, Escol attempted to start industries at Sellers Landing that did not depend on the steamboat market. In the 1870s, he dabbled in making coal gas and paper, though this necessitated new infusions of capital. When he went to his Philadelphia backers, he found them disgruntled that they had reaped little return on their initial investment.[115]

Twain and Warner dashed any further hope Escol had of retaining credibility. The residents of Sellers Landing, who experienced the limitations of his venture firsthand, bought into the notion that he was the prototype for the blundering Colonel Sellers in *The Gilded Age*. They helped spread the rumor. When it got back to Philadelphia, his chances of remaining solvent in Illinois died.[116]

In the 1880s, Escol settled in Tennessee, where he had recently consulted for railroads.[117] There, he wrote his account of his own and other engineers' triumphs, leaving out his work in the South and his rail-road system. He never did escape the embarrassment of Colonel Sellers. In 1892, Twain revived the character for his novel *The American Claimant*, in which the crazed inventor seeks to materialize the dead in a quest to create a workforce of zombies, among other ridiculous fancies.[118] Escol would live until the beginning of 1899, long enough to witness the entire eleven-year run of his memoirs in *American Machinist*. Even in his obituary, though, he did not escape the Colonel. Characteristic of the many papers that reported his passing, the *Atlanta Constitution* noted, "It has been said that he was the original of Mark Twain's 'Mulberry Sellers,' but he indignantly denied that imputation and resented any allusion to such being the facts."[119]

Escol's divergent experiences in his early years in the East, including his trip to Britain, and in his later work in the West and South, re-flected the uneven geography of the early and mid-nineteenth century. They supported Twain and Warner's cynical interpretation of industrial capitalism as well as his own very different version of the story. In the South, the participation of Escol and Charles in the slave economy and in highly speculative infrastructure schemes represented some of the great moral compromises and vain decisions northern capitalists made in search of personal gain. In the West, despite the pitfalls and obstacles of business in the early years, mechanics did build systems of transportation, manufacturing, and communication that resulted in a dense landscape of industrial cities above the Mason-Dixon Line. Through travel, learning, working, and forming continental and transatlantic networks of people, business, institutions, information, infrastructure, and other things, they mediated the rapid, tumultuous industrialization and urbanization of the early and mid-nineteenth century.

———————————— ✕ ————————————

CHAPTER 3

RATIONALIZING THE
FACTORY AND CITY

In December 1855, Coleman received his invitation to return to Philadelphia. His second cousins William and John Jr. recalled that when they last corresponded, "you were very desirous of obtaining a situation east of the mountains." They proposed that he take charge of the drawing office at their machine tool works, to replace their brother-in-law and partner Edward Bancroft, who had recently met an early death. Coleman would play a central role in the design of machines and organization of production at William Sellers & Co. The firm's namesake would continue to oversee mechanical operations on the shop floor as well as marketing and customer relations. They hoped this would remedy William's problem of "too little time to look after the execution of the work or to systematize the details so as to cheapen the construction."[1]

Coleman would be his second cousin's "scientific collaborateur" for over three decades, a partnership that spanned more than a few institutions and resembled in many ways the experiences of Nathan Sellers and his brothers and cousins.[2] Established in 1848 as Bancroft & Sellers, the firm had become William Sellers & Co. upon Bancroft's untimely death. Already one of the foremost machine tool builders in the East, under Coleman the firm turned the drafting room and shop floor into laboratories decades before the rise of corporate labs. This positioned them to influence industry on a broad scale.[3]

Like Coleman, William rarely stopped working and tinkering, mostly with machines. Both men participated in engineering societies and institutions of higher education, though William was also active in corporate and political organizations. Following his father John, William married into a prominent family of manufacturers in Wilmington, Delaware, wedding Quaker silversmith Ziba Ferris's daughter Mary in 1849. They had three children in the 1850s, Katherine, William, and a baby who died at age one. A serious young man, William was just thirty-one when he hired Coleman in 1855.

In the history of engineering, William Sellers is best known as the father of industrial standardization in America, for introducing in the 1860s what became the American standard screw thread and later a standard for the world. He promoted it through the Franklin Institute, which he and Coleman made the nation's leading technical society between the 1860s and 1880s, and through allied public and private institutions. Standard screw threads and other innovations at their firm made the most basic parts of machines and factories interchangeable, which helped integrate technological systems and industrial economies on a grand scale. Rationalization and standardization gave the Sellerses and other manufacturers the ability to set wages as well as prices, strengthening their sway over labor markets and the market for their products.[4]

Defining the terms of production proved important for eastern firms and cities seeking to maintain their competitive position in an expanding and increasingly diverse nation. In midcentury Philadelphia, economic development came to mean, as William Sellers's state senator and later congressman, William "Pig Iron" Kelley, proclaimed in 1852, that their city "ought to be the manufacturer for the increasing millions of luxurious citizens of her sister sovereignties in the South and West, and . . . dispute the manufacturing supremacy of England."[5] Firms like Sellers & Co., with numerous patents and markets in America and abroad, supported visions like these.[6] They were part of a group of interrelated machinery and metal firms that made Philadelphia the nation's leader in the sectors that supplied equipment to the factories and railroad networks of industrializing America. Equipping other factories to systematize production in a great range of sectors, they had pervasive effects on labor and business practices.

The project of rationalization remade the material life and economic geography of cities through technological as well as political interventions. The massive machine tools and power transmission systems built by Sellers & Co. formed the mechanical foundations of industrial districts, rail networks, and steamships that in turn structured the expansion and integration of regional economies.[7] The political economy of urban and industrial development in many northern cities supported this; and the year before William hired Coleman, Philadelphia capitalists were the first to consolidate their city with a vast area of surrounding factory towns, farmland, and emerging bedroom communities. This gave industrial interests new authority to regulate the working classes and to shape the region's urban and economic development.

Building the Factory City

In a series of papers titled "American Supremacy in Applied Mechanics" published at the end of the nineteenth century, Coleman Sellers asserted that "the growth of the machine tool industry . . . is a necessity . . . lying at the bottom of mechanical progress in supplying the means to accomplish not only its own production but all the machinery that is required for each and every industry in the world."[8] Because they made machines with which other manufacturers produced other machines and commodities, the midcentury emergence of firms like Sellers & Co., Pratt & Whitney in Hartford, and Brown & Sharpe in Providence had profound effects upon American industrialization.[9] Their products and the production they enabled built out the factory districts and cities of the first industrial revolution United States. Sellers & Co.'s early sales, plant, and customer base thus offer a detailed window into the development of the sectors and places they helped build.

In establishing and growing their business, Edward Bancroft and William Sellers paid constant attention to some of the most basic technological challenges facing manufacturers. As mechanics virtually from birth, they were genuinely interested in big engineering problems, which also gave them a chance to reach broad markets. Their early innovations were most striking in the area of power transmission. While in Providence, around 1845, Edward Bancroft invented and patented a new type of self-adjusting hanger upon which iron shafting (also known as millwork) could swivel to reach and provide power to machines even as they shifted position in a factory. This hanger had greater strength and durability than others on the market, and required less skilled labor to set up and adjust.[10] New England manufacturers initially declined to purchase it, however. This apparently led Edward to join his brother-in-law in Philadelphia in 1848, where William had returned the previous year and was making and repairing machines in West Philadelphia.

Although it was located close to the Pennsylvania Railroad's main Philadelphia workshops as well as many coal yards and lumberyards, the space William had rented was evidently too small to accommodate their venture. Its distance from the main machine-making centers of the region also limited their access to customers, suppliers, and skilled workers. Soon after Edward's arrival they moved to a factory in the district of Kensington, where their neighbors included steam-powered textile mills, iron foundries, rail depots, and shipbuilders along the Delaware River.

With a capital of $16,000 and exclusive patent rights, the firm started out on a more secure footing than most machine makers of the era.[11] They hired William's younger brother, John Sellers Jr., who had been employed making iron castings in a small shop since his return from Cincinnati. His experience in banking and his close family ties made him a logical choice as the company's bookkeeper.[12] Edward, William, and John Jr. all surely spent time on the shop floor, but according to engineering historian Joseph Roe, who interviewed Coleman in the early twentieth century, "Bancroft was the inventive member," designing tools while William oversaw production.[13]

Shafting was a fundamental element of factory systems in the nineteenth century, serving the same function that electrical wires do today in transmitting power to machines. Nineteenth-century British scientist Andrew Ure called millwork "the grand nerves and arteries which transmit vitality and volition, so to speak, with due steadiness, delicacy and speed to the automatic organs" of the factory.[14] It not only distributed power, but it also subdivided the aggregate power generated by a water wheel or steam engine into parcels of energy to run each machine in a factory at different speeds.[15] Necessary for any up-to-date workshop, shafting was usually a big investment. As Sellers & Co. noted in a catalog in the 1870s, "In any large factory the shafting . . ., considered as a machine to transmit motion, is most frequently the largest in the establishment."[16]

Bancroft & Sellers designed, produced, and installed their millwork systems—which included shafts, hangers, couplings, and pulleys—to drive all sorts of machines, from printing presses to the saws of lumber mills and the looms of textile factories. For their neighbors in Kensington, they put up lines of shafting to link different rooms in mills and connect production in separate buildings. They devised early freight elevators, setting up "power hoisting machines arranged to be worked in such manner that the operator can ascend or descend at pleasure." For one of the larger textile makers, they entered into an equipment management contract of sorts, agreeing to provide shafting to "put up in the mill from time to time as they may require it."[17] Through jobs like these, they influenced how other manufacturers organized production in their factories.

Figure 7. William Sellers & Co. machine shop, c. 1895, showing a portion of the turning and drilling departments. Lathes on the floor are connected via belts to the shafting above. William Sellers & Co., *Illustrated Catalogue and General Description of Improved Machine Tools for Working Metal* (Philadelphia: J. B. Lippincott, 1895), 5.

The firm also made the basic machine tools used in metalworking. Outside of millwork, their principal products were lathes and planers, massive tools for cutting and shaping iron, brass, and other metals with which manufacturers produced everything from locomotives and iron ships to nuts, bolts, agricultural implements, and plumbing supplies. Compared to shafting, which was sold by the pound and provided a relatively steady stream of business, planers and lathes brought less regular but quite profitable orders. They typically cost between $500 and $2,000 per machine ($14,300 to $57,000 in 2010), depending on their size.

The company's first order book, one of few surviving records, offers an in-depth view of its early business.[18] In it were recorded the specifications for machinery, arrangements for payment and delivery, and other notes regarding 626 orders received between 1848 and 1854. Compared with later company history, these records reveal rather unsystematic beginnings.

Early contracts were negotiated case by case, as the company had yet to work out standards for production schedules, pricing, and payment.

The firm's first big orders were lathes and a planer for local iron foundry Tasker & Morris and engine makers I. P. Morris and Merrick & Towne. In the first quarter of 1848, the firm also took orders from outside Philadelphia, from a Pittsburgh firm and from railroad car and papermaking machinery maker Pusey & Jones in Wilmington.[19] These were some of the region's leading metal and machine firms, suggesting that Edward and William benefited immediately from their families' relationships among manufacturers.

Their business quickly grew in scale and scope. By the fall of 1849, more than half of the orders were for shafting, including large contracts to outfit entire plants. By 1851, the company's line of specialized machine tools was sufficiently broad to furnish in one large order for the Nashville Manufacturing Co. two types of lathes, two types of planers, and three sorts of large drills. For some customers, Bancroft & Sellers designed custom machine tools and parts. They entertained regular requests to make replacement parts, with customers sending broken parts to copy or whole machines to repair. Often they collaborated with I. P. Morris and other manufacturers in networks of production that resembled those of Coleman Sellers & Sons two decades earlier.

They continued the practice of building their business through institutional networks, marketing their products to other manufacturers especially through the Franklin Institute.[20] At the institute's exhibition in 1850, Edward won a second prize for a dynamometer designed by Providence mechanic C. F. James and manufactured in the firm's Kensington factory.[21] Bancroft & Sellers won a silver medal for a planer at the institute's 1854 fair. They also participated in fairs in other regions, receiving a silver medal at the Maryland Institute's 1857 exhibition in Baltimore.[22]

Like other businesses in this era, Bancroft & Sellers relied on cousins and in-laws for orders, referrals, and access to information. When Escol's grade-climbing locomotives were installed at the Carbon Run Improvement Co., Bancroft & Sellers secured a contract to make shafting for the company. Samuel and Joseph Bancroft were regular customers and apparently referred their Brandywine neighbors. John Sellers, William and John Jr.'s father, did the same in Delaware County. Some orders from Wilmington likely resulted from connections made through William's in-laws, the Ferrises, as well as the Bancrofts and Pooles. Another customer was Sellers & Pennock, which branched out from riveted fire hose to make leather belting for machinery, which complemented Bancroft & Sellers's product line as it connected shafting to machines. This likely resulted in referrals and collaboration between the two firms.[23]

Edward and William kept up a lively correspondence with Coleman in Cincinnati well before his return to Philadelphia, marketing their tools to Niles & Co. and other western concerns.[24] They also traveled to promote their products directly to other manufacturers. After one such trip in August 1850, they recorded in the order book that nineteen "parties ordered samples of our patent ball & socket hangers & pillow blocks, sent to them, they having examined the model of the same as carried by our [Edward Bancroft] on his late trip Eastward."[25] These prospective customers included most of the leading machine makers in the Northeast. The firm maintained links with customers in more distant markets through brokers and commission merchants, who operated clearinghouses for machinery and accounted for a significant number of orders.

Through these networks, Bancroft & Sellers quickly developed a national market and reputation. They shipped planers and lathes to Ohio and Tennessee and shafting to Connecticut, Virginia, and San Francisco. The superior character of their work is suggested by regular orders from other makers of machine tools and shafting and from the U.S. Mint, where their cousin Franklin Peale still worked and which remained among the most advanced metallurgical establishments on the planet. They also sent their employees to other regions. In 1852, the Kershaw Steam Mill in Camden, South Carolina, ordered "all the necessary iron work for a saw mill." Bancroft & Sellers were to "furnish a competent workman to put up said work in the Steam Mill Co. paying his expenses out and back and his board whilst putting up the work." This was a routine service provided to customers, from Kensington to Georgia.

In Philadelphia alone, Bancroft & Sellers contributed to the growth of a wide range of sectors, including other firms with sizable impacts on the region's economic development. Their precision tools made possible the construction of high-speed engines for steamships, enabling Delaware River shipyards and allied engine builders to transform overseas and coastal shipping.[26] One customer made steam engines for the anthracite mines, blast furnaces, and rolling mills of eastern Pennsylvania that supplied the city with coal and iron. Another was Henry Disston's saw works in Kensington, which would become the largest saw maker in the nation. Other customers included scientific instrument makers, jewelers, lamp and gas fixture manufacturers, textile mills, lumber mills, sugar refiners, printers, publishers, and chemical and pharmaceutical plants, and the U.S. Navy Yard on the Delaware River.

The enterprises that integrated these manufacturers into a national marketplace, the railroads, represented a key portion of Bancroft & Sellers's customer base, as did local firms that served the railroads, such as Asa Whitney & Sons' wheel works and locomotive makers Baldwin and Norris. Bancroft & Sellers supplied machine tools and shafting to Mid-Atlantic and southern railroads, including some affiliated with Charles and Escol's associate Duff Green. Merchants in New York filed large orders for a railroad farther south, the Copiapo & Caldera line of Chile, in South America.

But Bancroft & Sellers's most important client of all was a line based in Philadelphia. On January 14, 1850, on a referral from Matthew Baird of the Baldwin Locomotive Works, Pennsylvania Railroad engineer (and later president) J. Edgar Thomson ordered an eight-foot lathe. This initiated a long relationship between William Sellers and the company that in the post–Civil War era would become the largest corporation in the world. Beginning the following year, Pennsylvania Railroad superintendents and master mechanics Herman Haupt, Enoch Lewis, and Strickland Kneass repeatedly visited the Kensington works to order shafting, lathes, planers, drills, bolt-making machines, circular saws, and later steam hammers, turntables, and cranes to outfit the railroad's locomotive, engine-building, and repair shops. These shops sustained the network that came to dominate the nation's industrial heartland.[27]

The railroads defined the economic geography of the nineteenth-century cities and their industries.[28] Running along the edges of Philadelphia's commercial and financial downtown, rail corridors became manufacturing corridors as firms—especially those engaged in heavy production—sought access to coal and iron shipments, shipping service and freight depots, and the port. From their shops on Beach Street in Kensington, Bancroft & Sellers enjoyed easy connections to nearby depots of the Philadelphia & Reading and North Penn railroads, and to the Camden & Atlantic Railroad wharves and Charleston and Savannah steamship landings along the river. Other neighbors included lumberyards, coal yards, and customers such as the Kensington Iron & Steel Works and shipbuilder Reaney & Neafie. To the north, the tracks of the North Penn ran through the region's largest concentration of textile mills to the district of Port Richmond, where they connected to the Reading's immense coal wharves and the nearby I. P. Morris foundry. Shipping merchants and commission brokers located their offices close to these depots.

Traveling west from Kensington, the Reading line that skirted the northern edge of the downtown linked up with the Germantown &

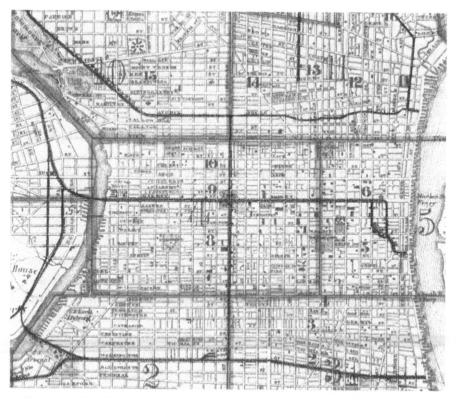

Figure 8. Railroad lines encircling central Philadelphia and connecting manufacturing districts at the outskirts of the mid-nineteenth-century city, adapted from Samuel Smedley's *Atlas of the City of Philadelphia,* 1862, plate index. Map Collection, Free Library of Philadelphia.

Norristown Railroad at Ninth Street, where warehouses and locomotive shops located together with other Bancroft & Sellers customers in the chemical and printing sectors. Farther west, the Reading tracks met the City Railroad at Broad Street, the site of more depots and a cluster of large machine and metal manufacturers, mostly serving the railroads, in a district known as Bush Hill. The City Railroad ran south down Broad Street and connected with the Pennsylvania Railroad depots, the U.S. Mint, and the West Chester & Philadelphia Railroad depot along Market Street. Farther south, they met the Philadelphia, Wilmington & Baltimore line running along the southern edge of the central city. This last railroad led east to depots and wharves on the Delaware, where it gave access to the navy yard and the Southwark metalworking district, home to Merrick & Towne and Tasker & Morris. These lines formed the skeleton of Philadelphia manufacturing's geography.

For specialized makers of heavy machines, the connections provided by the railroads and the agglomeration of firms in these districts enabled production and access to markets.[29] Although these lines did not all run on a standard gauge track by midcentury, without the network of rails, I. P. Morris in Port Richmond and Bancroft & Sellers in Kensington would have struggled to collaborate in the fabrication of large tools for the U.S. Mint on Market Street, let alone reach customers farther away. As employers of blacksmiths and mechanics located together in these districts, they attracted pools of skilled labor from which they could draw for their fluctuating workforce needs. After five years in business, this, together with a need for more space, led Bancroft & Sellers to move west along the Reading line to Bush Hill.

By the 1850s, Bush Hill was one of the major centers of machine building in the world. Bancroft & Sellers's new neighbors included the Baldwin and Norris locomotive plants, Asa Whitney's car wheel works, machine maker Rush & Muhlenberg, fellow tool builder Bement & Dougherty, and Hoopes & Townsend, owned by William's old schoolmate Barton Hoopes, which would later become one of the largest producers of nuts and bolts in the world. These metalworkers turned out an array of complementary products, largely for the expanding railroad market. Baldwin, Norris, and Whitney purchased tools from Bancroft & Sellers and Bement & Dougherty. Whitney supplied wheels to the locomotive makers. Hoopes & Townsend provided a steady supply of nuts, bolts, and screws. All enjoyed low freight charges and strong networks of word-of-mouth labor recruiting among their workers, who mostly lived nearby.[30] Politically, the proprietors and mechanics of Bush Hill made up a voting bloc whose interests were represented in Harrisburg and later Washington by Pig Iron Kelley.

In developing their new plant on an entire block at Sixteenth and Hamilton Streets, Bancroft & Sellers used the most common form of financing for manufacturers, the mortgage. A merchant and banker named Hartman Kuhn and his partner Henry Beckett effectively invested in the firm by deeding the block to Edward and William in return for a rent due every six months. A clause in the deed enabled Bancroft & Sellers to purchase the property outright when they had accumulated enough capital. To protect their investment, Kuhn and Beckett stipulated that the firm must erect "substantial buildings of sufficient value to secure the . . . yearly rent."[31] In the ensuing decade, the firm would buy up the block to the north, doubling the size of its plant.[32] Edward purchased a four-story brick house around the corner, and William moved his residence from Kensington to the area of Bush Hill, too.[33]

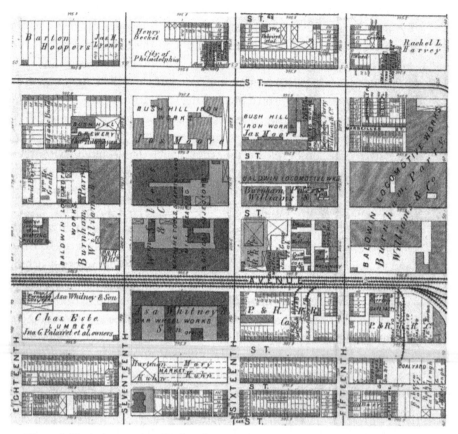

Figure 9. Fire insurance atlas of Bush Hill in 1875, showing its large plants along the railroad and beyond them the homes of workers, managers, and firm proprietors. From: G. M. Hopkins, City Atlas of Philadelphia, vol. 6, Wards 2 through 20, 29, and 30, plate L (Philadelphia, 1875). Map Collection, Free Library of Philadelphia.

They designed their new factory from scratch, accommodating a foundry, machine shop, and pattern shops in one-story buildings whose roofs drew the heat off the shop floor. Offices and drawing rooms occupied an adjacent two-story building. A spur from the railroad ran through the entire plant.[34] Ultimately, the machine shop would include railway turntables where locomotives and freight cars could turn to deliver or collect machines and supplies. An engine built by I. P. Morris drove the machinery, fed with coal from underground vaults.

Industrial boosters celebrated the plant's architecture and equipment. Charles Robson marveled at the "very perfect system of cranes" that moved heavy materials.[35] In *The Royal Road to Wealth*, I. L. Vansant lauded the machine shop for its "great solidity . . . and consequent freedom from

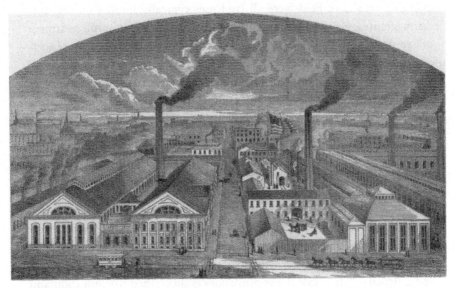

Figure 10. Lithograph of William Sellers & Co. factory, c. 1868, with neighboring factories of Bush Hill and in the distance the church steeples and smokestacks of other districts. From I. L. Vansant, ed. *The Royal Road to Wealth, an Illustrated History of the Successful Business Houses of Philadelphia* (Philadelphia: Loag, c. 1868), 4.

vibration. Heavy walls, with continuous iron plates, form the floor for the smaller tools, lathes, and planing machines—while all large tools are erected on substantial foundations of their own."[36] With large windows, wrote Robson, "the whole space is admirably lighted and well ventilated, while, in winter, all parts of their buildings are heated by steam."[37] As drawings in the boosters' tracts illustrated, with these sorts of investments, smokestacks soon supplanted church spires as the dominant features of Philadelphia's skyline.

Most factory districts in Philadelphia County, however, were not yet part of the city proper in 1853, lying instead just outside its borders. Bush Hill was situated in the district of Spring Garden, one of twenty-nine different municipalities that made up the county. By midcentury, the area had been transformed from a group of large farms and country estates to a burgeoning neighborhood of row homes and factories along the railroad, which the state had laid out in the 1830s.

In addition to the railroad, at least two secondary factors led to Bush Hill's development as a manufacturing district. First, a large number of undesirable institutions compromised the area's allure as a site for elite housing. The State Penitentiary, House of Refuge, City Hospital, Blind Institution, Magdalen Asylum, Orphan Asylum, Widow's Asylum, and city gasworks tank field were all land uses typically located on the out-skirts of cities. One nearby institution, the Wills Asylum for the treatment of eye injuries, was particularly well suited to address the results of accidents common to factories.[38]

The second dynamic that encouraged industrial development was an active growth politics among the commissioners who governed Spring Garden. Historians have characterized the public sector of the county's outlying townships and districts as underdeveloped.[39] But Spring Garden had its own board of health, guardians of the poor, and a joint police force with several other townships, as well as water, gas, and public school services shared with the city. As the district's population grew from 3,498 in 1820 to near 60,000 by midcentury, the commissioners expended considerable tax revenue in leveling hills, burying creeks, building wharves on the Schuylkill River, and paving wide thoroughfares "to afford sites for market houses, and to give a freer passage to increasing currents of trade, and railroads."[40] In 1851, the commissioners subscribed to eight thousand shares of the Pennsylvania Railroad on the condition that the railroad—which did not pass through the district—would guarantee uniform prices for delivery throughout the county.[41]

The commissioners bolstered these investments with a long list of ordinances regulating everything from bread and oysters to horses, dogs, hogs, goats, and vaccinations; from hay and straw to turpentine and guns; and from railroads and markets to the construction of buildings. An entire class of laws addressed nuisance industries and their by-products, including distillers, soap boilers, slaughterhouses, and disposal of coal ash.[42] But, significantly, they passed no noteworthy restrictions on metalworking, except for the prohibition of work on Sunday. Indeed, some of the commissioners, including Eli Kirk Price, were major investors in Spring Garden foundries.

Like Price, Philadelphia County's largest property owner, Hartman Kuhn was part of a class of real estate owners who played an important role in developing factory districts. A former director of the Second Bank of the United States, by the time of his mortgage to Bancroft & Sellers in 1853, at age sixty-nine, he had amassed considerable holdings around Philadelphia County.[43] Some were in areas where residential neighborhoods would grow, but others were in emerging factory districts.[44] Though there is no evidence he put capital directly into manufacturing firms, as Philadelphia bankers rarely did, the mortgages he offered amounted to a crucial investment in developing the region's industrial geography. In this way, real estate interests were part of what political scientists call a "growth machine."[45] In the case of mid-nineteenth-century industrial districts, this took the form of a loose network of manufacturers, railroads, financiers, and allied politicians and boosters.

As suggested by the Spring Garden commissioners' investment in the Pennsylvania Railroad, the growth politics of the district operated within a broader geography that included the county and the state legislature. In the fall of 1853, Philadelphians elected Eli K. Price and Bush Hill locomotive builder Matthias Baldwin to the legislature on a platform of consolidating the various districts, boroughs, and townships of the county into one vast municipality. This initiative, which passed the following year, had been brewing for more than a decade. It was partly a response to working-class riots in the 1840s, including one against a railroad under construction. It also presented capitalists an opportunity to exercise greater control over economic development.[46]

According to Price, consolidation would "enable Philadelphia to sustain the rivalry of other cities seeking to absorb her trade, wealth and population."[47] A single city government presiding over more than 130 square miles promised more efficient coordination of transportation, schooling, and urban growth, including the development of railroads and manufacturing districts. The city's elite could now move out of the crowded downtown, away from factory smokestacks and the poor, while retaining the rights of municipal citizenship and influencing the business climate.

There is no evidence that William and Edward were active in the movement for consolidation, though the Sellerses would participate in and benefit from the alliance of capitalists that formed as a result. Historian Andrew Heath argues that consolidation spurred the region's bourgeoisie to define its interests as a more unified class of mercantile and industrial capital.[48] When the bill passed, the commercial newspaper the *Price List* proclaimed, "the merchant feels that his own strength, in giving direction to public opinion and action in question touching his own or the common welfare, will be most virtual by equal union with the manufacturer and mechanic."[49]

Consolidation inspired a wave of heightened industrial boosterism. William Sellers and Samuel Merrick were the first to aid Edwin Freedley in compiling his tome, *Philadelphia and Its Manufactures*. They gave Freedley access to the region's workshops and garnered subscriptions among such colleagues as Matthias Baldwin, Bement & Dougherty, and I. P. Morris. In response, Freedley promoted Sellers, Bement, and their machines with particular vigor, proclaiming, "Philadelphia Tools unquestionably surpass those made elsewhere in this country, in *strength*, *proportion,* and *workmanship.*" They were "average nearly or quite double the weight allowed by other American builders," making them more solid and durable.[50] Sellers and Bement, he asserted, "have produced machines that may fairly be regarded as mechanical triumphs; and have given a permanent reputation to the manufacture which will make Philadelphia, if not already fairly entitled to be so called, the great seat of this business in the United States."[51]

He ended his tract by quoting the president of the city's commodities exchange, who described the sensual experience of "our Steam Engines . . . plying their iron arms in every street. In every by-way is heard the sound of the shuttle and the clink of the hammer." Freedley cast factories as monuments and celebrated the heightened metabolism they brought to the urban economy and environment, declaring "many a stately edifice, with its hundreds of employees and clanging machinery, sends forth a stirring music to quicken the pulse of our city life."[52]

Despite the boosters' inflated language, their arguments reflected Philadelphia manufacturers' understanding of their own place in the national and world economy. Outsiders validated their view. The national credit rating firm R. G. Dun repeatedly called Bancroft & Sellers "the best tool makers in the U.S.," with "splendid shops."[53] By the 1850s, local machine tool makers—Sellers and Bement foremost among them—had built the largest concentration of this sector in the country.[54] Their tools and shafting systems played a key role in America's antebellum manufacturing boom, and they laid the mechanical foundations of Philadelphia's industrial vitality.

However, machines alone did not make industry efficient and cities competitive. The significance of Sellers & Co.'s work lay not just in the things they made, but also in their impacts on the organization of production and innovation, a critical yet often ignored part of economic development. If product sales reflected the firm's present value and contribution to the region's economy, innovation signaled its ability to spur future growth.

Organizing Innovation

At midcentury most factories still operated on a decidedly unsystematic basis, including to some extent Sellers & Co. Coleman's arrival heralded a turning point in this regard. The Dun agency estimated the firm's capital at the time at $100,000 ($2.6 million in 2010).[55] Sellers & Co. invested some of this wealth in what would become a continuous pattern of innovation that transformed most parts of the plant as well as much of American and

European manufacturing practice. In addition to their own experiments and internal reorganization, Sellers & Co. collaborated with other engineers, scientists, and firms, making innovation a networked phenomenon.

In mechanical engineering, innovation depended upon rationalization. According to historian John K. Brown, this meant that the firm "sought to establish an ordered body of engineering knowledge, a systematic approach to innovation, and structured controls over production . . . creating and imposing order on disparate materials, men, machines, and methods."[56] Rationalization took several forms at Sellers & Co., most importantly experiments to improve and standardize their shafting and render the individual parts of machines interchangeable. This expanded their product line with new technologies for diverse customers. Through detailed cost accounting and the overhaul of their drawing office, they also instituted systematic reforms in their management procedures, including piece-rates of payment for most operations on the shop floor.

In its early years, the firm had followed the custom of selling millwork by the pound. This provided incentives to make a lighter product—as opposed to a better product—to compete in the marketplace. Sellers & Co.'s later *Treatise on Machine Tools* recounted the turning point. "In 1856, feeling that this system . . . was not the proper way to dispose of such things . . . we instituted an extensive series of experiments to demonstrate just how strong, and consequently how heavy, each article . . . should be."[57] The standard couplings developed by the firm employed half as much iron, permitted the use of stronger hangers to attach shafting to the ceiling, and sold for 25 percent less than conventional couplings. They could be mounted without skilled labor.[58]

Selling shafting by the foot, Sellers & Co. reconfigured the relationship between manufacturers and their customers. As the trade journal *American Machinist* later noted, "Under the old system, or want of system, the purchaser was ignorant of the total cost until the completion of the work; whereas under the new system it is always possible to ascertain the cost before the order is given to execute the work."[59] The firm became a dominant price-setter in the American shafting and machine tool markets.

With these standards of design and pricing established, William and Coleman reformed systems of "mechanical execution" on the shop floor. They developed standard gauges and special machine tools "to insure, under proper system of inspection, work of the best quality, in the largest quantities, without the use of workmen of great skill."[60] The company's millwork and its constituent parts gained a measure of interchangeability previously unattained in this area of manufacture. This, in turn, reduced the cost of labor.

These changes enabled the firm to undersell competitors, who were ultimately forced to adopt most of the Sellers & Co. systems. Between 1858 and 1871, they sold over 13,400 couplings. *American Machinist* reported that "the newly-erected cotton and woolen mills" in post–Civil War New England "alone necessitated so large a production that often the firm of William Sellers & Co. had as much as 10 miles of shafting under construction at one time."[61]

Implementing these innovations brought changes in production on the shop floor that evidently upset labor relations. Some hint of workers' reactions comes in the 1870s from a reporter who interviewed William, whose veiled reference to "some close struggles with ignorance and prejudice" suggests resistance to reduced discretion over their craft.[62] Virtually every element of rationalization increased employers' power over labor, as managers came to dictate small details of work. The firm's machines and management innovations made major contributions to manufacturers' authority over their workers.

However, Sellers & Co. itself, like Baldwin and other specialized machine builders, avoided momentous conflict with labor. The nature of work and the high wages at the firm help explain why. Like their miller and mechanic ancestors, and like many nineteenth-century machine builders, William and Coleman based their professional identities and their company's fortunes on what Brown has termed "a policy of continuous innovation." This gave them strong incentives to preserve a high level of skill among their workers.[63] Employees making machine tools retained considerable autonomy on the shop floor, and their work remained more mentally challenging and less repetitive than other manufacturing labor. Sellers & Co. operated an internal apprenticeship program, and experienced no strikes or lockouts during the nineteenth century.[64]

Employment and wages at the firm grew across the second half of the nineteenth century. The workforce jumped from 190 to 423 men between the census years of 1860 and 1870, while output rose from $205,000 to $707,542 (machine tools were roughly 40 percent of total output, while shafting and, later, steam injectors for locomotive engines made up the rest).[65] Over the next decade, the firm's payroll nearly doubled again, as workers toiled in separately managed departments making tools, steam injectors, shafting, and railway turntables.[66] In the 1870s, the company began to pay workers on a piecework system that gave them both a measure of control over the pace of work as well as incentives to earn more through increased output and innovations of their own.[67] By 1880, according to the U.S. Census, machinists at Sellers made an average of $2.00 a day, blacksmiths made $2.20, and draftsmen made substantially more— at a time when the typical take-home pay among industrial workers was about $1.00 a day.[68]

Much of the company's managerial reform was organized from John Jr.'s bookkeeping department and especially the drafting office run by Coleman. During the latter half of the 1850s, they amassed meticulous data on the cost of producing each part of every machine tool sold, allowing them to set prices that typically garnered about 20 percent profit.[69] Detailed drawings provided instructions to founders, smiths, and machinists, subdividing and standardizing most tasks.[70] The extensive time and capital expended in growing the drafting office in this decade formed what Brown calls "an institutional foundation for long-term leadership in the machine tool industry."[71]

Edward Bancroft's son, John Sellers Bancroft, who spent most of his career at the company, called "the drawing room . . . a bureau of record of the judgment and conclusions of the practical men who design and who make the machine."[72] Creating and preserving drawings that were used as the basis for repeated alterations and improvements, it was the central repository of intellectual property. The drafting office developed and applied algebraic equations to sort out the dimensions and proportions of all parts of a machine, effectively codifying much of the design process itself.[73] It was this approach that led Edwin Freedley to declare in an 1856 treatise, *Leading Pursuits and Leading Men* (of the United States, with a decidedly Philadelphia bent), that "mathematical—not proximate— accuracy is their standard," and it earned William and Coleman a reputation as "scientific designers."[74]

The firm's design department operated at the leading edge of professionalization among mechanical draftsmen, who were previously drawn from the ranks of artists and architectural draftsmen. Sellers & Co. attracted some of the industry's leading talent. Before coming to work under Coleman, William Thorne directed his own machine tool company. Wilfred Lewis joined the department upon his graduation from MIT, and he went on to become the leading authority on gearing in the United States. Theodore Bergner, who began his career as a photograph painter, would later become a prominent consultant specializing in brewery machinery.[75] These men were longtime members of the American Society of Mechanical Engineers, the national professional society that counted William Sellers among its founders and Coleman among its earliest presidents.[76]

Complementing its innovations in mechanical drawing, the firm explored the new medium of photography. Coleman later recounted (in the third person) his earliest application of this emerging technology to the work of the firm, which in 1858, "desiring to illustrate their machinery by means of photography, engaged the services of portrait photographers for the purpose, with very poor results." So, "despairing of getting the work done by then professional operators," he "decided to learn the art himself."[77]

After experiments and consultation with others using photographic technology, in 1861 Coleman patented a predecessor of the motion picture projector called the " 'kinematoscope,' by which stereoscopic photographs of moving objects could be exhibited."[78] "What I aim to accomplish," he wrote in his application for the patent, is to "exhibit stereoscopic pictures as to make them represent objects in motion such as the revolving wheels of machinery, and various motions of the human body, adding . . . a semblance of life that can only come from motion."[79]

In the same year, Coleman helped establish the New York–based Amateur Photographic Exchange Club. Out of this national association, the following year he founded the Photographic Society of Philadelphia, which counted several Franklin Institute colleagues, including his cousin Titian Peale, among its early members.[80] From Boston, Oliver Wendell Holmes, who presumably found the new art useful in his medical studies, initiated a correspondence with Coleman that would last three decades.[81]

Coleman became an active correspondent for the *British Journal of Photography*. According to one memorialist, he established himself "as *the* representative of American photography on the international . . . scene."[82]

Figure 11. A still frame from a motion picture Coleman took in 1859, with his toddler son Horace at the left and eldest son Coleman Jr., age six or seven, at the right, in their father's office at Sellers & Co. Coleman Jr. is hammering on a piece of metal. This screen capture was taken from a video posted on YouTube: http://www.youtube.com/watch?v=U ZawjJTLFns&feature=youtube_gdata_ player.

This transatlantic exchange took on an added dimension when his colleague Theodore Ely of the Pennsylvania Railroad took a "detective camera" to England "to study the railroads of foreign lands."[83]

Photography influenced industry inside and outside the factory. Coleman later noted, photography "has become a means of copying drawings in all large and many small machine shops. An engraving may flatter in describing a machine. A photograph shows us the thing as it is, and is next in value to seeing the workmanship itself."[84] Sellers & Co. hence employed photography in marketing. Coleman also supervised the establishment of a photographic studio at the neighboring Baldwin Locomotive Works and offered courses in the new art at the Franklin Institute.[85]

The Sellerses' and their employees' long list of innovations helped the firm and its managers secure over one hundred U.S. patents between 1857 and 1901, and more in Europe, for everything from planers and lathes to steam injectors, boilers, cranes, railway turntables, hydraulic pressure systems, steam hammers, riveters, and ordnance. William and other managers traveled to Europe and other parts of America seeking new technologies and customers, returning with ideas they refined through shop floor experiments and elaborated in the drafting office.[86] They also purchased the American patent rights to European inventions, playing important roles in the transatlantic exchange of machine and later steel technologies.

Among William's most significant patents were those for injectors that supplied water to the boilers of locomotives and other steam engines. As recounted by longtime employees Alexander Outerbridge Jr. and Coleman Sellers Jr., "During a visit to England, in 1860, the attention of Mr. Sellers was called by Sharp, Stewart & Co., of Manchester, to the Giffard Injector for feeding steam boilers, a model of which had been sent by Flaud & Cie of Paris for the purpose of interesting English manufacturers." Upon his return, "a special department devoted to the manufacture was added to the plant."[87]

Sellers & Co. purchased a license from the eminent French engineer Henri Giffard to become the sole North American maker of his injector. The firm, by its own account, "spared no expense in investigations tending towards [its] further perfection."[88] Most notably, William patented an improvement that made his injectors automatically adjustable, doing away with the need to adjust water and steam pressure by hand and doing away with an entire apparatus of pumps. In 1861, the firm dispatched Escol Sellers to visit the Pennsylvania Railroad's shops at Altoona "for the purpose of observing the action of the Injector on the various Locomotives to which it is applied . . . to obtain such information as will Enable us to adopt the very best method" of its attachment to locomotive boilers.[89] Coleman collected extensive data on the performance of Sellers's injectors using various sources of water on the Reading and Lehigh Valley Railroads.[90]

Injectors became another large and pervasive product line for Sellers & Co. In his 1868 *History of American Manufactures*, Edwin Freedley reported the firm had "now forty hands constantly employed in manufacturing them, and their orders are at the present time fully up to their capacity for producing them."[91] The Pennsylvania Railroad made the Sellers injector a standard feature of all engines on its lines. Manufactories ranging from paper mills to iron furnaces adopted it, as well.[92] In 1878, *American Machinist* reported that innovation in steam injection was crossing the Atlantic from America back to Europe, as the French government ordered "drawings of injectors after the American model for use in the French marine service."[93]

By the 1880s, Sellers & Co. boasted an "experimental department established to eradicate defects of construction and to obtain a more complete development of the principle" of steam injection.[94] The head of this research laboratory was Strickland L. Kneass, the son of the Pennsylvania Railroad engineer of the same name, who took out some twenty-five injector patents himself. By the turn of the twentieth century his injector department included its own apprenticeship program to educate young mechanics and an exhibit room for marketing to potential customers.[95]

In addition to shafting and injectors, Sellers & Co.'s construction of new machine tools served the growth of the shipbuilding, locomotive, and iron and steel industries. On the same trip to England during which he encountered the Giffard injector, William visited the Newcastle plant of Robert Morrison & Co. and came away with the U.S. production rights for the Morrison Steam Hammer. Engineers at Sixteenth and Hamilton Streets adapted that as well. The hammer allowed metalworkers to make larger forgings than previously possible, especially at big iron and steel plants. Coleman called it "a necessity" for production of ironclad steamships.[96]

The firm's catalog enumerated the impacts of its tools, many of which enabled an ever-larger scale of production. Its sliding transfer tables and turntables for railroads and street railways—up to fifty-four feet in diameter—found customers "over the greater part of both continents of America." Locomotive and ship builders used Sellers & Co.'s pioneering hydraulic pressure systems for hammers, cranes, and other tools, as the company controlled "many of the most valuable patents covering the

Figure 12. A steam hammer built by William Sellers & Co. in the 1890s. William Sellers & Co., *Illustrated Catalogue and General Description of Improved Machine Tools for Working Metal* (Philadelphia: J. B. Lippincott, 1895), 245.

application of this power to riveting and working metals." These systems "tended to materially decrease the first cost of the plant" for a variety of large manufacturers, as "a very considerable number of machines or hoists may be operated from the same source of power with less cost for each machine, and very much less cost of conducting the power to it, than by any other known method." Sellers & Co. engineers calculated the production gains and long-term cost savings that their tools afforded customers, demonstrating their economic benefits.[97]

Among the company's foremost innovations was the development of interchangeable parts for all sorts of machines.[98] In an essay at the end of the nineteenth century, William elaborated on the significance of his 1857 patent machine to cut identical nut and bolt threads. "The attendant is no longer of necessity a machinist," he wrote, "for his only occupation is to provide his machine with bars, to remove its product, and to keep it clean, duties which attendance upon a number of such machines does not make onerous."[99] Fewer, less-skilled workers could now tend more machines turning out interchangeable parts. By 1860, the tool was selling in Europe. The firm's 1877 *Treatise on Machine-Tools* claimed, "One man has, in a day of ten hours, cut three thousand" screws on this machine.[100] Booster Charles Robson announced that the tool was "fast finding its way into every well-arranged workshop."[101]

Tools capable of turning out uniform products with less room for variation in workmanship revolutionized labor and productivity on the shop floor. Sellers & Co. articulated their motives in more than simple terms of saving labor, though many of their machines allowed managers to gain greater control over the actions and output of workers. The Sellerses themselves stressed the opportunity to deploy high-skilled labor in the service of greater innovation. The "problem of the day is not only how to secure more good workmen, but how to enable such workmen as are at our command to do good work, and how to enable the many really skillful mechanics to accomplish more and better work than heretofore," they asserted in the 1877 treatise.[102] The goal, at least in factories like theirs that required skilled labor, was not to de-skill the manufacturing workforce, but rather to enhance profitability and productive capacity.

This extended to the relationships between producers and purchasers. With standardized, interchangeable parts and detailed records contained in their drawing offices, firms would no longer need to expend valuable time, labor, and shipping costs in sending workers, tools, or broken parts to one another's factories. Underlying interchangeability were strict standards of production, or tolerances, that dictated just how far a tool's part could stray from the original, or reference standards. Sellers & Co.'s plant was known for the utmost commitment to precision. "Any one who will visit the establishment," Edwin Freedley averred, "cannot fail to be astonished at the extreme pains taken to insure accuracy in all parts of the machines which they make. . . . A variation of a hair's breadth, if it can be overcome, is not left unremedied."[103] The company sold a complete line of gauges, ensuring that customers could apply strict tolerances in their use of each tool.

Sellers & Co. promoted these standards of production and the machines that made them possible through shop visits, circulars, and treatises.[104] Sharing some results of their research, the firm's publications offered recommendations for workshop organization. In a discussion of their drill grinding machine, for instance, they wrote, "Any well-regulated shop should have a tool room, in which all the small tools required by workmen, such as drills, reamers, taps, mandrels, gauges, etc., are kept, and can be given out to the workmen to use, to be returned when done with or when needing sharpening." The drill grinder was fundamental equipment for this system, which "will make much fewer tools necessary to carry on business than will be required by the more shiftless way of allowing each workman to keep all the tools needed on his special work."[105]

By setting standards of shop practice, the firm reinforced customers' dependence upon their tools and disseminated new products and standards of efficiency among a wide range of companies and sectors. Innovation was the foundation of prosperity at Sixteenth and Hamilton, as at other specialized metal firms in Bush Hill and other districts in the city. This helped maintain Philadelphia's prominent place in nineteenth-century industry. Leading producers that set national standards in business and technology helped reinforce the city's status as a manufacturing center of the first order. They further institutionalized this position and the dynamics of innovation that sustained that status through trade and technical societies. The most important of these was the Franklin Institute.

Engineering Institutional Influence

When William and Coleman returned to Philadelphia at midcentury, the Franklin Institute was in serious trouble. The depression of the 1830s decimated its finances, a crisis compounded by debt it amassed when it bought the city's Masonic Hall for new exhibition space. As it focused more on science than industry in the 1840s and '50s, the establishment of the better-funded Smithsonian Institution and the American Association for the Advancement of Science challenged its position on the national scene. In 1857, as part of a movement to refocus the institute on "concrete industrial issues" and build on the strength of the region's machinery and railroad companies, William won election to its board of managers.[106] With his machine-building colleagues he redirected the institute's mission and saved it from financial ruin. They proceeded to revitalize its role in the regional economy and make it an important complement to their firms' research, product development, and influence in American manufacturing.

William's early service on the board focused on stabilizing the institute's finances. Although attendance at its annual fair reached one hundred thousand in the 1850s, the institute's money problems remained acute through 1863. At the height of the Civil War, a friendly takeover installed forty-year-old William as its president and overhauled the board's membership. Other than William and Coleman, only three of the old managers remained. The new generation on the board included shipbuilder Charles Cramp, toolmaker James Dougherty, Charles Parry from Baldwin, and J. Vaughan Merrick, son of institute founder Samuel V. Merrick.[107] Their firms were all busy with war contracts, which helps explain how, the following year, William and J. V. Merrick eliminated the institute's debt by garnering donations from twenty-seven colleagues.[108] They also changed its constitution and by-laws to allow the sale of stock, which raised further capital.[109]

The institute gained a sound financial footing largely since its managers resuscitated its relevance to the region's manufacturers. They turned it into a more formal and professional engineering society focused on the technical concerns of their businesses, two decades before the American Society of Mechanical Engineers was established.[110] A committee headed by Merrick and including Coleman proposed that membership meetings abandon administrative matters in favor of "reading a paper descriptive of some valuable improvement in the Arts, some alleged discovery in the Science, or great work in Engineering proposed, or in progress of construction, in this Country." They recommended that "inventors and others should be invited to bring objects of interest to the meetings." This opportunity to "draw forth desultory remarks from practical men, unused to speaking in public," would make the institute's meetings more inviting to businessmen and engineers. It would also revive its role as a forum promoting innovation and the dissemination of new technologies.[111]

The institute's leaders worked to strengthen its relationships with firms and institutions within and outside the region. Its membership grew as Philadelphia's major machine builders, including Baldwin, Bement & Dougherty, and Merrick & Sons, involved their employees in its activities. This effectively professionalized their workforces as they frequented its library and mechanical drafting courses. Sellers employees who joined in the 1860s included machinists like nephew J. Sellers Bancroft, molders, blacksmiths, sales managers, and accountant and cousin Charles W. Peale. Engineers from Sellers & Co. and other firms published articles in the institute's journal.[112] William's personal secretary at the institute, Henry Morton, served as its lecturer in mechanics, while also teaching chemistry and physics at the University of Pennsylvania.

The managers reached out to organizations that complemented and elevated the work of their firms. They hosted the American Iron & Steel Association's machine for testing the strength and properties of metals in the institute's building on Seventh Street, reestablishing a commitment to collaborative research. In the spring of 1864, following the explosion of the USS *Princeton*, they joined with the National Academy of Science in a Navy Department investigation on the expansion of steam. This was a question of central consequence to the speed, safety, and odds of victory of the gunboats presently battling Confederate troops.[113] When William went to Europe when the war ended the following year, the managers adopted a formal resolution recognizing the "opportunity for making . . . more intimate relations with the Correspondents of the Institute and with Scientific institutions and men of Science."[114] However, their most influential project initiated in the 1860s, and that for which William is remembered by historians and engineers today, was the development and promotion of a standard system of screw threads.[115]

The ridges that run around the shaft of a screw or bolt may seem tiny and mundane in comparison to the locomotives, steamships, and power tools that reshaped the geography and material life of the nineteenth century. But they were the most basic elements that held together those larger machines. "If there is any one thing in the transactions of the machine shop more incomprehensible than another," wrote the editor of *Scientific American* in 1863, "it is the want of some settled size or number for screw threads."[116] Addressing this problem, on the evening of April 21, 1864, at the regular monthly meeting of the Franklin Institute, William read an essay, "A Standard System of Screw Threads," that touched off a regional, national, and ultimately international movement to standardize the most fundamental elements and practices of industry.

"The importance of a uniform system of screw threads and nuts is so generally acknowledged by the engineering profession," he asserted, "that it needs no argument to set forth its advantages." The fact that large technological systems such as railroads that together spanned the globe were not mechanically integrated with one another was retarding the development of national and international markets, including those for his machines. "In this country no organized attempt has as yet been made to establish any system," Sellers continued, "each manufacturer having adopted whatever his judgment may have dictated as the best, or as most convenient for himself; but the importance of the works now in progress, and the extent to which manufacturing has attained, admonish us that so radical a defect should be allowed to exist no longer."[117]

The wartime economy heightened this sense of urgency. Local factories were at work on large federal orders for munitions, ships, uniforms, blankets, and machines to supply the Union army and navy. The city's railroads, primarily the Pennsylvania and the Philadelphia, Wilmington & Baltimore line, were the main transporters of troops and supplies along the northern front. But these roads did not use the same standards, right down to the bolts and screws in their train cars. This lack of interchangeability also characterized the machinery of steamboats that plied rivers and oceans, the streetcars and gasworks of the nation's burgeoning cities, and the engines and tools that drove factories around the world. Each manufacturer, railroad, or steamship line had to produce its own screws or order custom replacement parts according to their varying specifications for the pitch (the angle) of threads and the number of threads per inch. This often resulted in costly delays in production or service.

In his own business, William had long confronted the challenges born of this lack of a standard system, and he had worked with his customers to remedy this. A decade earlier, in a single order for the Pennsylvania Railroad, he had furnished screws with eleven different categories of threads per inch.[118] By the 1860s, he helped the railroad apply a uniform standard in its workshops, supplying templates and gauges to follow the system he proposed that night at the institute.[119] He thus unified the mechanical practices of the railroad's network dispersed across hundreds of miles, allowing workers in its West Philadelphia shops, for example, to routinely replace parts on locomotives built in faraway Altoona.

By 1864 most of the major metalworking firms in Philadelphia used Sellers & Co. tools, including William's bolt maker, so they had a certain investment in promoting his system as the national standard.[120] Another longtime customer and ally was the Eagle Screw Co. of Providence, whose mass production of pointed screws helped make it the largest screw maker in the United States.[121] By the time he presented his paper, William had persuaded other major manufacturers in the East to adopt the system. This was important for its chances of wider acceptance, since it did have competitors.

The Sellers system did not emerge in a technological vacuum. The eminent British engineer Sir Joseph Whitworth had already developed and promoted a standard, used in some American shops. The principal features of his system were a fifty-five-degree angle of threads and a rounded top to his screws. William proposed a sixty-degree angle and flat tops for both screws and bolts, arguing that this system was easier to make with uniformity, required workmen to make fewer cuts with fewer machines, and would thus cost less for manufacturers to use. He suggested a simple equation for determining the depth of the grooves between each thread, which was based on extensive testing at his tool works.

Beyond any advantages inherent to the design, it was the Franklin Institute and its coordination of an entire community of engineers and industrialists—and its networking with other institutions—that made possible the dissemination of the Sellers standard. When William finished reading his paper that night, J. V. Merrick "submitted for examination specimens of a series of screw threads, bolts, and nuts, as used by" Merrick & Sons. Bement & Dougherty showed "drawings of bolts and nuts as used in their establishment." The assistant superintendent of the Reading Railroad and Baldwin superintendent Charles Parry shared the proportions used by their companies. On motion of John Towne the institute resolved "to send a copy of the paper read this evening . . . to the American Institute, New York, and to other similar societies, with the request that they will give it an early consideration with a view to promote the introduction into general use of the system advocated."[122]

A second resolution that evening appointed "a committee to investigate the question of a proper system of screw threads, bolt-heads, and nuts, to be recommended by this Institute, for the general adoption by American engineers."[123] Headed by Sellers's neighbor and competitor William Bement, it also included Merrick, Towne, two engineers from Baldwin, and his cousin Coleman. They tested the Sellers and Whitworth systems in the plant of Bement & Dougherty. Given the commercial advantages of having their region dictate the national standard, the committee predictably returned with a strong endorsement of the Sellers system.

In an effort to guide corporate and government technology policy and thereby grow the markets for Philadelphia machines, the institute and its members launched a decades-long campaign to spread the system. They appealed to the business sense of American engineers and manufacturers, stressing the impact on their profits. Bement wrote to the editor of *Scientific American*, inviting the opinions of "all good *practical* mechanics"—as opposed to academics or amateur technologists who might prolong the debate. The magazine congratulated American engineers that they had not succumbed to "the toils of schemers and theorists who would have confused instead of making the subject plain and practical."[124] As historian Philip Scranton has noted, "Resisters, caricatured as backward and foolish, understood that standardization was about power and money, not just progress and rationality."[125]

The institute's leaders took special pains to lobby for the system's adoption by national institutions. They sent formal resolutions to the army quartermaster general, the chiefs of ordnance of the army and navy, the chief engineer of the military railroad corps, and the superintendents and master mechanics of railroad companies and major machine shops. These missives called for "requiring all builders under any new contracts to conform to the proportions recommended."[126] William and his colleagues recognized that these institutions had the collective power to dictate the terms of machine production nationally.

The U.S. Navy took the lead among public institutions in adopting the Sellers standard. In 1868, its Bureau of Steam Engineering established an investigative Board to Recommend a Standard Gauge for Bolts, Nuts and Screws Threads. This effort apparently favored Sellers by design, as the board's chair was Philadelphia Naval Yard engineer Theodore Zeller, who had close personal and contracting relationships with the region's machine builders. The institute's collaboration on federal technical investigations and Navy Department engineers' tradition of publishing in its journal also helped maintain longstanding ties. After touring eastern machine shops, many of which used Sellers & Co. tools, the board reported that his thread enjoyed the "very important advantage of ease of production" over the Whitworth system.[127] Since the navy had become the chief mechanical authority in government, its endorsement effectively made the Sellers standard the federal standard.

The Pennsylvania Railroad, arguably the most influential company in American industry, officially endorsed the system in 1869 and subsequently branded itself as the "standard railroad" of the nation. Three years later the professional association for railcar and railway mechanics, the Master Car-Builders' Association, followed suit.[128] As railroads and railcar makers adopted the standard, the machine shops with which they contracted were forced to employ it as well. In an 1874 address to railroad engineers in the Master Mechanics Association, Coleman Sellers proclaimed, "the primary object of this Association may almost be said to [be to] introduce uniformity in all parts of the great railroad system of the United States."[129] A decade later, just twenty years after William initiated this "organized attempt," there was hardly a North American railroad that did not employ his standard.

Spreading the system must have figured in the Sellerses' motivation to help establish national publications and institutions of engineers. "Believing it to be of the highest importance that such a standard should be generally recognized as soon as possible," announced the editors of *American Machinist* in their first issue, in 1877, "we heartily second all efforts looking to that end."[130] The American Society of Mechanical Engineers, which originated in the offices of *American Machinist* and included William and Coleman among its earliest promoters, took up the cause at its first annual meeting in 1880.[131] When one of the society's founders, Alexander Holley, was preparing entries on twelve hundred engineering terms for *Webster's Dictionary*, it was Coleman who, guiding him around the shops of Sellers & Co., helped define the terms related to screws.[132]

Engineers pursued and debated standardization of various things (and still do), almost always with considerable implications for the prospects of a particular region's or nation's industries.[133] Historians have emphasized that the Sellers system of screw threads spread slowly and was resisted by many manufacturers as they debated and promoted a variety of industrial standards in the post–Civil War era. Scranton notes, "Its adoption would force costly adjustments by firms . . . changes that could loosen . . . the ties of dependency between machinery builders and their clients without obvious benefit to the former."[134] In the 1870s and '80s, American engineers also battled one another over the prospect of adopting the metric system. Franklin Institute members led the charge against it. Coleman published a series of articles in defense of the English system.[135] The Sellerses also promoted standard machinery, pushing tools they patented and produced.[136] Coleman Jr. argued for a "Franklin Institute Standard of Mechanical Drawing."[137]

Although adoption of the Sellers system of screw threads was certainly piecemeal and uneven, and decades passed before it became predominant among American manufacturers, it ultimately became a worldwide standard. It spread across the United States with the help of tools made by Sellers & Co., of course, but also by New England's leading tool builders Pratt & Whitney and Brown & Sharpe, which developed standard gauging tools for the system.[138] After extensive correspondence with European governments and engineering societies, the International Congress for the standardization of screw threads endorsed the system at Zurich in 1898.[139] A half-century later, the British finally adopted it, as well.

Though highly technical in their details, the projects of rationalization and especially standardization were economic development strategies, ones in which engineers played the central roles and effected profound changes in patterns of work and material life. The systems of machinery, production, and management pioneered at Sellers & Co. in the mid-nineteenth century helped produce a new geography of factories, railroads, and cities. From screw threads, to the standard time zones developed to coordinate railroad scheduling, to the regularization of track widths at the end of the century, the economic and physical integration of the industrial metropolis and nation depended upon the capacity of machines, technological systems, and people to work together.

In their firm, William and Coleman organized innovation in ways that helped make Philadelphia the leading center of high-skilled manufacturing in America. This work and the profits it generated would be the foundation of William's substantial influence in the region and in national industry during the second half of the nineteenth century. Standardization and municipal consolidation were two episodes in a larger set of strategies that manufacturers and fellow capitalists pursued to build and exercise power over urban and economic development. These pursuits and the alliances they forged gave them vital roles in the Civil War and in postwar metropolitan growth.

CHAPTER 4

✕

PROGRESSIVE ECONOMIC
DEVELOPMENT

By the time of the Civil War, William Sellers was an influential leader among Philadelphia capitalists at large, not just manufacturers. Perhaps the best evidence of this was his membership in the Saturday Club. From the 1860s to '80s, this exclusive group included the chief power brokers of the region's elite Republican set. Investment bankers Anthony Drexel and Clarence Clark and merchant John Welsh, later U.S. ambassador to England, were national figures in finance and trade. Pennsylvania Railroad president Thomas Scott and machine builders William Bement and Sellers represented industry. Board of Trade president and longtime Franklin Institute board secretary Frederick Fraley was an important bridge between mercantile and industrial capital. Intellectual leadership sat at the table in the persons of political economist Henry Carey and University of Pennsylvania provosts Charles Stillé and William Pepper. Other public opinion makers in the club included publishers J. B. Lippincott, Henry Charles Lea, and Morton McMichael, who was mayor of Philadelphia just after the war.[1]

These men met once a week for a dinner of oysters, truffled turkey, beef à la mode, ices (though the menu stipulated "no confectionary"), and cigars.[2] The more important item on the menu was their deliberation about what they viewed as the most pressing issues facing the region. They cultivated national connections by routinely inviting such leaders as New York mayor Abram Hewitt and President-elect Chester Arthur.[3] But the

Figure 13a. Portrait of William Sellers, c. 1880. Thomas Scharf and Thompson Westcott, *History of Philadelphia, 1609–1884*, vol. 3 (Philadelphia: J. H. Everts, 1884), facing 2264.

Figure 13b. Portrait of Coleman Sellers, c. 1895. Edward Dean Adams, *Niagara Power History of the Niagara Falls Power Company, 1886–1918*, vol. 1, *History and Power Projects* (Niagara Falls, NY: Niagara Falls Power Co., 1927), 362.

club's main purpose was to define the agenda for local civic affairs and economic development.

According to one member, the Saturday Club "fairly represented the working force of the city, and gave to this force a unity which neither New York nor Boston possessed."[4] Yet this claim masked what they were up against, and indeed the very need for them to coordinate in such a concerted fashion. During and after the Civil War, the Sellerses and their colleagues formed networks of firms and institutions as they competed with another class of capitalists who gained increasing power in the city and region. These were the bosses of the city's political machine, the more popular wing of the local and state Republican Party. Their wealth derived from real estate, oil, and monopolies over transit and utilities.

Their influence grew as they incorporated the city's expanding immigrant communities into their constituency, threatening the power of the elite, including its ability to shape economic development.[5]

Positioning themselves against the machine, the Saturday Club's members pushed visions of public life that illuminate the origins of Progressive urban reform. Historians date the Progressive Era to the 1890s and subsequent decades.[6] However, the social and government reforms promoted by the various institutions that the Sellerses and their allies led beginning in the 1860s illustrate the deeper history of Progressivism. The Saturday Club was part of a larger complex of institutions that the Sellerses and their allies built or rebuilt during and after the Civil War. Their collective work would give rise to the urban professions and disciplines, including urban planning, social work, public administration, and sociology, around the turn of the twentieth century. The intellectual and institutional roots of the new sciences of urbanism were thus based to a great degree in the economic development aims of postwar industrial capital.

From the Civil War until their deaths in the first decade of the twentieth century, William, Coleman, and John Sellers Jr. held prominent positions of corporate, professional, and institutional leadership. During the war, they made their tool works a major military contractor. William and John Jr. joined in founding the city's Union League. After the war, they grew their portfolio of businesses, establishing the Edge Moor Iron Co. to make structural steel in 1868 and acquiring a controlling interest in the Midvale Steel Works in 1873. Sellers & Co. and Edge Moor remained family firms, while Midvale was a joint venture of financiers, manufacturers, and railroads.

The Sellerses' postwar institutional pursuits amplified their influence over economic development within and well beyond Philadelphia. As directors of institutions including the University of Pennsylvania, the Pennsylvania Museum and School of Industrial Art, and the Franklin Institute, they shaped technical education and labor markets. Through such forums as the Institute of Mechanical Engineers of Great Britain, of which William was a member, and the American Society of Mechanical Engineers, they helped structure communication among manufacturers and scientists around the world. William played a central role in the Centennial Exhibition of 1876, the world's fair that boosted industrial Philadelphia's claim to the title "Workshop of the World."

If the region's early nineteenth-century complex of institutions fostered its transition to an industrial economy and society, the set of institutions

that grew up during and after the war established a more focused and ultimately professionalized paradigm of economic development. Universities,
technical institutes, and urban reform became an increasingly important
part of cities' competition with one another. Industrialists like the Sellerses
tied the work of firms and institutions together, more concertedly and
coherently than earlier generations, helping drive this change.

Like their counterparts in other cities, the Sellerses and fellow Philadelphia capitalists invested considerable time, money, and ideas in the
institutional infrastructure of their corporate, educational, and social reform sectors. This complex remains visible in the civic sector and economic development of Philadelphia and other cities in the twenty-first
century. From the public schools to universities, educational institutions
complemented the work of the Franklin Institute and of individual manufacturers. These institutions organized research, supplied skilled labor,
and maintained connections to engineers and technical societies around
the globe.[7] Today, economic development scholars call the city's universities, teaching hospitals, and major cultural organizations, most of which
grew up in this era, "anchor institutions," signaling their ability to generate broader economic activity. The foundations of much of the wealth
and many of the relationships that built these institutions were laid in the
Civil War.

Union of Capital

Philadelphia won the war. At least that is how the city's capitalists viewed
the conflict. Financiers Jay Cooke and Anthony Drexel floated the war
bonds that bankrolled the Union victory. The North also won the war
in the workshop and on the railroad, thanks in large part to locally based
firms. The war effort helped unite Philadelphia capitalists and positioned
them to pursue a coordinated economic development agenda.

Production at Sellers and other firms shifted to military contracts during the war.[8] Local shipyards built ironclads. Textile mills made wool
blankets and blue uniforms. Kensington, Southwark, and Bush Hill metalworkers turned out ordnance for an increasingly industrial army and navy.
Pennsylvania Railroad vice president Thomas Scott led the U.S. Military
Railroads in the war's first year. The Pennsylvania Railroad itself took
on huge contracts to carry soldiers and provisions to Gettysburg and the
Ohio and Mississippi Valleys. With this capital, after the war it would buy

up smaller lines that soon made it the dominant railroad of the nation's industrial heartland. One of these lines, the Philadelphia, Wilmington & Baltimore Railroad, connected the city to the nation's capital and markets to the south. After the war, William Sellers would join its board. This was one more signal of his membership in the most elite circles of the region's capitalists.

The coalition of manufacturers, lawyers, and financiers that achieved the city's consolidation in 1854 formed an important constituency for the Union. It gave birth to a local Republican Club in 1856 and a Union League in 1862.[9] Amid the financial panic of 1857 and the resulting depression, Pennsylvania Republicans, led by Henry Carey, pushed a high-tariff plank that helped them win municipal and statewide elections in 1858. The national party added this stance to its antislavery platform at its 1860 convention in Chicago.[10] Carey and his agenda enjoyed the support of such magnates as Pennsylvania Railroad president J. Edgar Thomson, iron maker Joseph Wharton, Matthias Baldwin, John Towne, Asa Whitney, and the Sellers brothers. The region's industrialists increasingly joined the Board of Trade in the late 1850s, uniting with their mercantile and financial counterparts in a strong lobbying bloc.[11] Throughout the next half century, the tariff would be fundamental to visions of Philadelphia manufacturers, merchants, and skilled workers of a unified national market protected against cheap foreign labor in an international marketplace.

The Sellerses' and fellow manufacturers' views of the Union reflected their beliefs about how an industrial society and economy should be organized. Like their father, William and John Jr. took a stand against slavery, though their position was more moderate, owing largely to their business interests. They basically followed the party line of elite Republicans, viewing their mode of "free" economic development as both more humane and more rewarding financially. Their most public stance of which evidence survives from the 1850s was an endorsement of the Republican Club's condemnation of Congress in its handling of Kansas, which asserted that the nation's leaders had failed "to prohibit in the Territories those twin relics of barbarism, Polygamy and Slavery."[12]

While northern industrialists supported free labor and free soil, when Confederates fired on Fort Sumter in April 1861 these industrialists still enjoyed strong ties to the South, especially its cotton and markets for manufactures. Southern demand for mechanics and machinery reached a crisis point as secession loomed. In January 1860, George Simpson of Haw River, North Carolina, penned a letter to Coleman Sellers alerting him to a job opening for a master machinist at the North Carolina Central Railroad, with a yearly salary of $2,500 and "a very elegant Dwelling furnished . . . to live in." "My advice," Simpson offered, "would be for you to get the recommendation of some distinguished Democrats," as the company's president was "a great Democrat."[13]

Coleman declined the invitation, but in December, as the conflict between North and South deepened, Simpson wrote back with a proposition of greater urgency and national consequence. "I see [South] Carolina is willing to appropriate $50,000 to be given to any firm or company who will forthwith establish in one of the Southern states a manufactory for the making of improved arms . . . & will contract for $50,000 worth of arms annually for 5 years." Simpson claimed, "As soon as S. Carolina goes out of the Union, we can get" North Carolina (Simpson's own state) "to pass an act to allow us to manufacture any kind of arms regardless of any patent & to protect us in that right & should we be sued [the effect would be] harmless." He proposed an alliance with Sellers & Co., in which he offered access to southern contracts. "If you would come on to my house," he suggested, "I would go with you & visit some of the Southern Legislatures and make contracts to furnish them arms."[14]

Such a lucrative offer aside, and although Sellers & Co. already carried on a significant business with southern railroads and mills, Coleman and his cousins had, in the words of historian John K. Brown, "so many reasons to reject it and essentially none to accept it."[15] Nonetheless, Philadelphia manufacturers' dependence on southern markets and resources made secession an unacceptable prospect in their eyes. The week before they received Simpson's contracting proposal, William Sellers, Samuel Merrick, Matthew Baird, Isaac P. Morris, Asa Whitney, and other machine builders participated in a mass meeting sponsored by the city council to express friendship toward the South. Several people in attendance proposed concessions such as terminating Pennsylvania's personal liberty legislation and enforcing the federal law on fugitive slaves. Although William valued his southern business, he was not willing to support this sort of compromise. Several months later he and Merrick joined fellow Republicans and antislavery Democrats in a demonstration in support of Fort Sumter.[16]

Philadelphia capitalists galvanized loyalty to the Union through mutual association and social pressure. Aiming to suppress the seething Copperhead sentiment in the city, in November 1862 William joined with fellow Republicans and Saturday Club members in establishing a Union Club. The following month, they made over the club into a full-blown

Union League. John Sellers Jr. described the atmosphere surrounding its founding as "a period of great depression among loyal citizens . . . in consequence of the slow progress that had been made by the Government in quelling the rebellion. Its primary object," he declared, "was to bring to bear the weight of moral and social influences upon our community to countervail the ill effects of secession sentiments then beginning to be more openly expressed by . . . the Copperhead element in our city."[17]

Meeting in members' homes, this new organization had, by its own later account, a profound effect upon the city's elite. Business and civic leaders "gathered under our roofs . . . the best intellectual elements . . . night after night," to discuss wartime strategy in politics and commerce. "The moral power which our association exerted upon our members and our guests, soon touched the popular heart, and strengthened, consolidated, and organized the patriotic sentiment of our people." Among Philadelphia capitalists, this "had a powerful influence on the social position of disloyal men," who "were shut up within their own small coteries."[18] This affected their business as well as social standing.

The Union League mobilized troops and material support for the war effort, as well. The region's Republican elite threw their financial, political, and manufacturing weight behind President Lincoln and General Grant. In 1863 and '64, according to John Jr., the League's Board of Publication "distributed upward of two million copies of pamphlets in the English and German languages," making the case for the Union. "Another committee was organized to promote enlistments in the army, which during the same two years raised as many as ten thousand men for the service of their country."[19]

Sellers & Co., Bement & Dougherty, Merrick & Sons, Asa Whitney, and the Baldwin Locomotive Works recruited volunteers from their workforces, which included many native speakers of German. They organized companies of soldiers where, according to historian Andrew Dawson, "discipline and hierarchy mirrored the workshop."[20] The Union League appointed John Jr., together with John Towne and Matthew Baird, to a committee "for the purpose of providing employment for Disabled Soldiers who have been honorably discharged from the United States service."[21]

In 1864, the Union League campaigned for Lincoln's reelection and, like groups in other cities, organized a fair to benefit the hospitals of the U.S. Sanitary Commission.[22] On Logan Square, across the street from William Sellers's residence at 1819 Vine Street, temporary pavilions housed a restaurant, a Turkish tobacco parlor, a German Club, and galleries for the display of fine arts, manufactures, and arms and trophies, including two smokestacks of the USS *Monitor* "perforated by Rebel shot."[23] To garner extra funds for this Great Central Fair, which ultimately took in over one million dollars, John Jr. hosted a soiree at his home. The evening's entertainment consisted of music, literary readings, and "parlor magic" by Coleman.[24]

In addition to raising troops and cash, the Sellerses devoted their mechanical and manufacturing energies to the Union, and profited in the process. In 1861, Coleman patented "a new and improved mode of making projectiles."[25] Sellers & Co. filled orders for cannons and thousands of shells for the local Frankford Arsenal.[26] Between 1862 and 1865, the firm furnished over one hundred machine tools for federal armories in Massachusetts, Philadelphia, Washington, and St. Louis, the navy yards of New York, Washington, and New Hampshire, as well as the U.S. Military Railroad. Total receipts for government contracts between 1862 and 1867 amounted to nearly $170,000 (over $2.4 million in 2010).[27] These jobs, together with orders from private railroads and other manufacturers engaged in military contracting, kept the works buzzing with activity.

The Sellerses' approach to the war was colored more by economic considerations than by radical politics. Although they supported the Emancipation Proclamation and African Americans in uniform, they were cautious not to confront the racial prejudice prevalent among their workers and fellow Philadelphians. Despite their Union League commitments, the hectic business climate led Sellers & Co. to discourage workers from volunteering and to delay sending their militia company off to fight. When Confederate forces pushed north into Pennsylvania, however, a flurry of volunteering and the draft sparked a labor crisis at Sixteenth and Hamilton Streets and other plants. In July 1863, Coleman wrote Escol, "Lee's raid has taken off to the war a company of our best men so you see we are in a bad fix—with lots of work and no men to do it with."[28]

Labor shortages aside, Sellers & Co. and fellow contractors such as Baldwin, Bement, and Cramp had a very profitable war, allowing them to grow and reorganize their shops in the postwar period.[29] The output of the city's machinery firms more than doubled from $3.1 million in 1860 to $6.5 million in 1866. That of iron products soared from $2.9 million to $13.1 million during the same period.[30] This translated into personal wealth for the Sellers and other manufacturers.

In an 1865 listing of the "Rich Men of Philadelphia" compiled from income tax data, William Sellers was among thirty-six residents—all of them bankers and manufacturers—with incomes over $100,000 (nearly $1.4 million in 2010).[31] That same year, partly in response to an attempt by I. P. Morris to lure away their chief engineer, Sellers & Co. raised Coleman's salary from $4,000 to $10,000, or 6 percent of the firm's net yearly profits.[32] R. G. Dun estimated Sellers & Co.'s capital at $200,000 to $300,000 in 1864, calling their credit "good every way, undoubted & strong." In 1873, when Coleman, J. Sellers Bancroft, and business manager James C. Brooks were admitted to the partnership, Dun reported that they "own large [real estate] . . . made a great deal of money & [are worth] over [$]1,000,000."[33] This figure reached $3 million in 1885. When the firm incorporated the following year—as a closely held concern whose $1.5 million in stock was all owned by William, John Jr., Coleman, and Bancroft—the agency reported a net worth, including the factory, machinery, and liquid assets, amounting to "[$]4,000,000 strong" (over $95 million in 2010).[34]

Despite his wealth, William remained a workaholic, almost a stereotypical engineer in his obsession with technical matters. Big changes occurred in his family life, as his first wife, Mary, died in 1870. Three years later he married Amélie Haasz, daughter of a Philadelphia grand piano maker.[35] Over the next decade they would have two sons, Alexander and Richard, and a daughter who died at age two. William and Amélie socialized with the families of other industrial magnates, such as the Whartons and the Rosengarten family of chemical manufacturers.

However, in letters to Amélie as she vacationed in exclusive Newport, Rhode Island, while he was home occupied with work, William professed feeling uncomfortable and awkward in society circles. He enjoyed fishing with colleagues out of the West Island Club near Newport, but found fox hunting distasteful. In a playful jab at his rough engineer's ways, he wrote her, "Some men like myself and some boys like mine . . . need a feminine mind to guide them[;] the trouble is however that we blurt out something that no amount of diplomatic skill can make good."[36]

Alone with their young children on holiday, Amélie complained, "I wish you had associated with you some man who carried as much weight as yourself, but they are all so little known and without monetary influence that it makes the brunt of the thing come upon you—and that should *not* be." Although William included his brother in all of his companies, John Jr. remained a minor player in all but internal management. "When a man reaches your position, and is president of all his ventures," Amélie continued, "he should have those under him to whom he can relegate his duty—work—while he quietly advised them *at his leisure*."[37] A few days later she reconsidered, admitting "my feelings here are very divided." The "people one brushes against" in Newport and other elite settings "seem to lead such empty lives"—even some close friends of William's, who "have built a barrier of exclusiveness based upon pride of birth or pride of wealth." Ultimately, she wrote, "I have come to the conclusion . . . that nothing in life would give you greater contentment than the continuance of your mechanical pursuits." He would be "miserable" otherwise.[38]

These sentiments notwithstanding, William led an increasingly public life in both corporate and institutional arenas during and after the war. Capital accumulation and alliances with other businessmen and politicians strengthened manufacturers' position in the regional and national economy. Close ties to the Treasury Department and the military bolstered their place in the Union. With a momentary power vacuum in the South after the war, industrialists from Philadelphia, New York, and Boston worked with their allies in Congress, the White House, and the military to influence fiscal, trade, and Reconstruction policy as well as federal contracting.[39] The Union League helped organize delegations of civic and business leaders from northern cities to visit Richmond and the South. But their greatest focus was on reforming Philadelphia and its complex of economic development institutions.

Networks of Reform and Development

With victory over the rebel states, Philadelphia financiers and manufacturers gained a new opportunity to capture markets in the South and overseas. But as they competed with rival cities and their industries, they also struggled to control the economic development of their own region. With the Republican machine ascendant in city politics, elite capitalists like the Sellerses launched a series of institutional efforts to retain their influence over local government and public affairs as well as the growing working classes. Manufacturers like William, Coleman, and John Jr. thus joined with fellow business and civic leaders out of necessity as well as opportunity. Their alliances sought to define what they viewed as the rational and therefore right ways to govern cities and the economy.

The vision they promoted of the public good included national policies that favored their industries and state and local reforms aimed at curtailing the power of the machine. The institutional vehicles for advancing

this agenda ranged from government commissions to nearly mass movements to the most private forums like the Saturday Club. This institutional
complex, the alliance of elite capitalists who led it, and the agenda they
advanced would define much of the later Progressive movement.

In his essay "American Supremacy in Applied Mechanics," Coleman
offered his own version of how the different groups of local capitalists
allied to form a unified class. "A machinist of Philadelphia who had made
for himself a competency and had settled down to enjoy the fruits of his
industry was made a director in an old and long-established banking
institution. He found to his surprise that neither farmers nor mechanics were recognized, but the tradesmen and merchants were the chief
customers of the institution." To make the connection, he began by inviting some bank officers to visit Bush Hill's locomotive and machine
works. Subsequently, the machinist-turned-bank-director "gave a dinner
to which he invited and seated at the table alternately, bankers, mechanics and merchants; this with the sole intention of correcting a fault which
as he thought existed in the community, and to advance the mechanic
arts."[40] This was one of many networking efforts in this era, and one of
the less institutionalized ones.

As Coleman recognized, the marriage of mercantile, financial, and
manufacturing capital allowed local firms to influence federal trade policy
and capture widespread markets for their products. Family capitalism persisted in late nineteenth-century cities, but the regional, national, and international industrial economies became more integrated as these groups
leveraged one another's assets and connections.[41] Historian Sven Beckert
makes a similar argument in explaining the formation of a metropolitan
business class in New York and its growing national power in the second
half of the nineteenth century.[42]

Philadelphia's elite capitalists united earlier than their New York counterparts, and their networks focused to some extent on leadership in different sectors. New York's most powerful economic institutions were in
finance and trade, while Chicago's Board of Trade and Mercantile Exchange protected its position as a center of commodity trading and processing in the West. Philadelphia's leading economic institutions were
more closely tied to engineering, high-tech manufacturing, and the diversity of industry it supported. Like their counterparts in other cities,
however, the Sellerses and other elite, socially conservative Republicans
battled to maintain a metropolitan political economy that advanced their
interests over those of the machine.

The Union League's membership included the broadest swath of Republican businessmen in the city. By 1873, it counted over eighteen hundred members. They ranged from machine politicians to elite reformists like the Sellerses, who made the association itself a battleground for competing interests. William and John Jr. envisioned the institution as a vehicle for electoral and civil service reform. To further this agenda, William donated $1,100 "to be expended in prizes for essays on the best mode of making nominations to office."[43] Reflecting on the league's history in 1901, John Jr. asserted, "It is its duty to be sensitive and influential in a dignified and forceful way in home politics by exerting its authority to maintain the sanctity of the ballot."[44] William charged that the machine was robbing Philadelphians of a representative government. "The primary election is so completely dominated by machine politics that few of us feel it necessary or even useful to attend it. At the general election fraudulent voting, and fraudulent counting of votes, have become notorious. Has the Union League no duty to perform when this treason is spreading in our midst?"[45]

Throughout the later nineteenth century, the Sellerses and their colleagues in the Saturday Club, and some of those in the Union League, battled with the "contractor bosses" and "utility monopolists" of the machine for control of local and state legislative bodies, executive offices, and the Republican Party. As the Sellerses' urgent messages at the turn of the century suggest, they enjoyed only limited success. In 1865, they managed to elect newspaper publisher and protectionist Morton McMichael to the mayoralty. However, in the words of historian Peter McCaffery, his administration "proved to be a short-lived triumph for respectability." Career politicians gained the upper hand in the party, and by the end of the decade "the struggle between reformers (respectability) and 'bosses' (corruption) had begun in earnest."[46]

City contracting, public works, and the gas, streetcar, and electric utilities were the province of bosses such as builder "Sunny Jim" McNichol, Standard Oil partner William Elkins, and butcher-turned-traction magnate Peter Widener. These men coalesced around political clubs including the All-Night Poker Players and the Society of Mysterious Pilgrims. In 1875, a watchdog committee of the Union League condemned this last group as a "dictatorial band of men, nominally of both parties, but without true allegiance to either, which now rules and oppresses our city and is disgracing and destroying the Republican organization."[47] The reformers attempted to maintain some measure of control over Philadelphia and its growth through such public and private institutions as the Fairmount Park Commission, the Pennsylvania Industrial League, and the Philadelphia Social Science Association. They took on electoral and governance problems through political reform associations, most of which were ephemeral.

From the late 1860s through the 1890s, William and Coleman Sellers joined the likes of McMichael, Henry C. Lea, Clarence Clark, J. Vaughan Merrick, and William Bement in a range of political reform movements. At various times, their efforts spawned the Citizens' Municipal Reform Association, the Reform Club, the Committee of 100, the Civil Service Reform Association, and the Citizens' Committee of '95 for Good City Government.[48] Coleman engaged in national networks of reformers through the American Institute of Civics and served as his neighborhood's ward captain for the Citizens' Municipal Reform Association.

William strategized with Lea and others to influence gubernatorial and presidential nominations, regain control of the state Republican Party, and prevent the lease of the city-owned gasworks to private companies tied to the machine.[49] The reformers won partial victories with a new state constitution in 1873 and the Bullitt Bill of 1886, which gave the city a new charter with more efficient organization of departments. The Sellerses remained "in entire sympathy with every effort to obtain the best men for office and to disconnect our civil service from politics."[50]

Yet William lost faith in the capacity of universal suffrage to foster responsible city government. Writing to an associate in 1891, he admitted, "The political corruption we desire to abate, is the legitimate outcome of our existing system." Philadelphia's city council must be composed of men "whose ambition and pecuniary interest will be, to have the best interests of the City guarded at all points, and we cannot have such men, upon the basis of universal suffrage." Rather, "It is the direct interest of a large, perhaps the largest proportion of the voters, to provide work for themselves or their friends, which other people must pay for." He called universal suffrage "indispensable for the proper government of a State, but a Municipality is a very different affair."[51]

A city, as William put it, "in fact, is a corporation, and under our present system, the property of the corporation is managed by parties who hold no stock in it."[52] Some, like the utility monopolists and their allies in real estate, perhaps held too much stock in the city's physical development. But unlike Sellers, they were not invested in sectors that supported economic development in a wide range of industries that enabled

Philadelphia to compete in the national and world economy. William and his colleagues thus spent much energy seeking alternatives to the way their rivals ran the city.

In November 1869, they established the Philadelphia Social Science Association (PSSA) as a branch of the American Social Science Association. The local contingent soon accounted for half of the national membership, composed mainly of lawyers, businessmen, and academics committed to applying the emerging social sciences to foster efficient government. They found an unofficial home at the University of Pennsylvania (Penn), where members included Provost Stillé, professors Lorin Blodgett and Edmund James, and trustees William Sellers, John Welsh, and association President Henry C. Lea. Their self-proclaimed purpose was "to guide the public to the best practical means of promoting the amendment of laws, the advancement of education, the prevention and repression of crime, the reformation of criminals, the furtherance of public morality, the adoption of sanitary regulations, and the diffusion of sound principles of economy, trade and finance."[53]

The PSSA's agenda represents a direct antecedent to—and fairly covered the scope of—the Progressive movement that would remake American cities and society in later decades. Drawing upon research by local and international scholars of law and public administration, the association promoted compulsory education, civil service and tax reform, and the arbitration of strikes. Papers read at its meetings compared Philadelphia to other cities in North America and Europe, often exploring best practices of poor relief, gasworks administration, and sometimes working-class housing.[54] The "useful results of these meetings," claimed the PSSA's annual report for 1873, "may be traced in the action of the legislature on questions of taxation," as well as in "various matters introduced into the new [state] constitution."[55] Later that decade, the organization assisted in the creation of a state Board of Health. A group of members organized the Free Library of Philadelphia.[56]

The university sought to promote systematic empirical research and dissemination of knowledge and what professions today call "best practices" related to the PSSA's agenda. The trustees tasked Edmund James with much of this when they appointed him as the first director of its new business school and professor of public finances and administration in 1883. Six years later, he organized the American Academy of Political and Social Science, dedicated to fostering social and economic policy discussion based on "scientific thought" from a broad range of disciplines.

The American Academy took the work of the PSSA to a national scale, linking academic and allied research to the emerging Progressive movement. In its first decade, it published papers on such topics as taxation, railway regulation, welfare, and settlement houses. James and his colleagues succeeded in making what began as largely an organization of people from Penn, with colleagues from Swarthmore and Bryn Mawr Colleges, into an organization with international membership and great prestige.[57]

Locally, James's research on municipal gas systems formed the empirical, scientific basis of the PSSA's battle against the utility monopolists of the machine. However, the bosses' United Gas Improvement holding company ultimately won a thirty-year lease of the Philadelphia Gas Works in 1897, the year following James's departure for the University of Chicago. The holding company would run the works into the 1970s.[58]

Beyond their mixed results at reforming governance and social policy, William Sellers and his colleagues were most active and most successful at influencing economic policy. A large contingent of Saturday Club, Union League, and PSSA members allied in the postwar period to lobby for state and federal measures friendly to the region's industries. Elite Republican manufacturers and politicians including "Pig Iron" Kelley—now a congressman—gathered in Henry Carey's weekly salon, the "Carey Vespers." Their guests included President Grant and former secretary of the treasury Salmon Chase.[59] Between the Carey Vespers, Saturday Club, and other meetings, William seems to have dined out most nights of the week during the 1870s.

Philadelphians pushed the high tariff through the Pennsylvania Industrial League, led by William, Morton McMichael, Joseph Wharton, and Henry C. Lea. This organization "lobbied in Washington camouflaged as a national American Industrial League" and maintained close ties to the American Iron and Steel Association, whose publication *Iron Age* voiced the league's policy concerns to the industry.[60] The league also influenced lawmakers in Harrisburg. William's eulogists claimed, "The Pennsylvania limited liability corporation law was enacted largely because of his efforts."[61] Sellers & Co.'s incorporation in 1886 apparently aimed to take advantage of this protection (it also may have helped the firm manage Coleman's retirement that year).

Through the mid-1880s, the Franklin Institute did its part in ensuring that financiers, publishers, insurance company directors, and attorneys joined its ranks. It kept them invested in engineering, and to a great extent in metalworking and chemicals as the core sectors driving innovation in regional manufacturing.[62] The Industrial League and the Iron and Steel Association split with Carey in 1875 over his support of a greenback monetary policy (which advocated paper currency over adherence to the gold standard).[63] But their members continued to promote the region's industry and its protection, including through the Manufacturers' Club and the Engineers' Club of Philadelphia. Coleman Sellers Jr. and a group of his younger colleagues founded the latter institution following the Centennial Exhibition of 1876, to promote closer social ties among a growing community of engineers.

The Centennial Exhibition itself was the boldest statement of elite Philadelphia capitalists' visions for their region on the national and global stage. It was also another strategy to transcend some of what they viewed as the limits of the machine's approach to economic development. Back in 1869, William had launched the effort to hold a world's fair in the city, persuading his Franklin Institute colleagues to memorialize Congress to that effect. They enlisted the aid of city council, financier Jay Cooke, and Pig Iron Kelley, who lobbied hard in the House.[64] Although Congress sanctioned the location of the nation's one hundredth birthday party in Philadelphia, it neglected to appropriate more than token funding. William, Frederick Fraley, and other members of the Saturday Club and Union League, of which William was vice president in the period 1870–74, mobilized local capital to carry it off. They did this largely through the Centennial Finance Committee headed by John Welsh and including Sellers, Fraley, Charles Stillé, and Matthew Baird.

The celebration was a major occasion to promote Philadelphia and its manufactures. Gearing up for the centennial, the Franklin Institute revived in 1874 its Exhibition of American Manufacturers, with funds from Sellers & Co., Asa Whitney, Baldwin Locomotive, Anthony Drexel, and others. At the closing ceremony, Coleman, then the institute's president, hailed the machines on display as the prime source of the reconstructed nation's prosperity. "These dumb servitors of mankind, with their frames of iron and their sinews of steel, which for life and motion devour the hard rocks of our anthracite, and in living and moving, breathe from their metallic lungs the hot vapors of steam," he declared, "are doing our work better, ministering to our wants more freely than could thousands upon thousands of slaves."[65]

Two years later, Philadelphia capital broadcast this message to a wider audience at the centennial. Machines and manufactures figured more prominently than at any previous world's fair. More than forty thousand exhibitors hailed from six continents. Local products gained not only the most advantageous exhibit spaces but also the highest accolades from international committees of judges. The committee on machine tools agreed that Sellers & Co.'s display of thirty-three massive specimens was "probably without a parallel in the past history of international exhibitions."[66]

At the close of the exhibition, a group of men including Baldwin Locomotive partner Charles Parry, publisher Charles Lippincott, and Pennsylvania Railroad engineer Strickland Kneass formed the International Exhibition Co. They purchased the main hall at the fairgrounds in Fairmount Park, then the largest building in the world, from the Board of Finance and negotiated with William Sellers and other members of the Fairmount Park Commission to keep it open.[67] Noting that the fair had served "to create a demand for many works of industry and fine art which had heretofore not found a market in this country," the company announced its intention "to continue the good effects derived from the Centennial, by affording unusually favorable facilities to manufacturers and producers to bring their wares to the further notice of the public."[68] Its circular depicted a map of Philadelphia's port, inviting businessmen to a global marketplace.

The pursuit of national and overseas markets was a primary goal of the men who had organized the Centennial Exhibition. Sellers & Co. invested a great deal in its exhibit in Machinery Hall and took full advantage of international opportunities afforded by the fair. In the summer of 1876, the visitors' log at its plant registered more entries than any other period in the company's history. Guests included engineers from Puerto Rico, Colombia, Spain, Belgium, and Austria; manufacturers from Scotland, France, Armenia, and Canada; scientists from England, Germany, and Italy; reporters from London and Poland; government representatives from Brazil and Australia; and visitors from practically every state or territory of the United States.[69]

The firm's sales and reputation expanded accordingly, aided by its Manhattan sales office (the Sellerses' Edge Moor Iron Co. ultimately counted sales representatives in New York, Detroit, Chicago, and Minneapolis). In 1878, Sellers & Co. began advertising in the national publication *American Machinist*. In its notes on American metal firms' sales, the journal reported that the firm had taken large orders for tools from South America, New Zealand, and New South Wales (Australia). "Their turn-table has been adopted on almost all railroads in North and South America and some have been sent to Europe."[70] Injector department head Strickland Kneass announced his products were employed on railroads "in England,

France, Russia, Belgium, Italy, Spain, India, China, Egypt, British Colonies, South America, West Indies, &c."[71]

Sellers & Co.'s markets became more national and global also thanks to their participation in exhibitions outside the region. The firm won high honors for displays at the 1867 World's Fair in Paris; the American Institute exhibition in New York in 1869; the World's Fair in Vienna in 1873; the exposition of the Imperial Technological Society in St. Petersburg, Russia, in 1880; the Exhibition of Railway Appliances in Chicago in 1883; the World's Fair in Paris in 1889; and at the 1904 Louisiana Purchase Exposition in St. Louis.[72] The thirty-five-thousand-pound planer exhibited by the firm at Paris in 1867 was used "to plane the massive bronze doors of the 'Palais de Justice,' and finally found its home in Holland," according to one Philadelphia booster.[73] European machine makers paid the company the highest compliment in their attempts to copy this and other Sellers machines, "many of which have been adopted as patterns by the constructors of tools in other countries," noted the international jury at Vienna.[74] William and his firms thus established a prominent reputation around the world, just as the cities that hosted these fairs and exhibitions reinforced their status as major nodes in the international industrial economy.

To a great extent, elite capitalists' ability to connect to national and world markets enabled them to sustain their allied projects of urban reform and development. Between the Civil War and the 1880s, Philadelphia's business and civic elite effectively positioned their city as a national and international leader in engineering, manufacturing, and railroads. A web of institutions supported their political, business, and social agendas at the local, state, and federal levels, helping them define the public interest in terms that favored their enterprises and promoted their visions of governance and civil society. In the set of problems they addressed, the institutions they built, and the ways they linked academia, policy, and civic reform, they helped give birth to the Progressive movement and the major changes it wrought in American cities. A large part of this work focused on education, an area with especially significant implications for their businesses and for economic development.

The Educational-Industrial Complex

In addition to practicing engineers and manufacturers, visitors to Sellers & Co. in the summer of 1876 included a teacher from Russia, a physics professor from Ohio, a university docent from Austria, a professor of mechanics from Norway, and the superintendent of schools of Tokyo.[75] Technical schools occupied a prominent place among American and Russian displays at the fairgrounds. But these guests made their way to the factory to observe its apprenticeship and research programs.

Both the visitors and their hosts recognized that technical education was not simply a matter of schooling. Historians of American education have portrayed late nineteenth- and early twentieth-century technical schooling as a narrow movement.[76] But in fact it was part of a much larger educational-industrial complex.

The Sellerses and their colleagues influenced the educational programs of public and private schools at all levels, shop-floor apprenticeships, and professional and technical societies. They often linked these institutions and firms in networked research and educational endeavors. The Social Science Association's effort to prevent strikes was just one part of this larger project to cultivate a workforce that operated as industrialists wanted. Elite manufacturers' support of technical education for a variety of workers, from mechanics on the shop floor to scientists in universities, was one of the key ways to keep their firms competitive in the high-skilled sectors responsible for much of the region's industrial prosperity in the late nineteenth century.

The Sellers firms were educational institutions in themselves, and they enjoyed close complementary relationships with teaching and research programs at the region's schools and technical societies. Sellers children and nephews attended college but went to Sixteenth and Hamilton Street for more practical education. John Jr.'s son Howard entered the drafting department upon his graduation in 1877 from Penn, where his uncle William had a large influence on engineering education. William and his first wife Mary's son, William Ferris Sellers, graduated the year before and entered the employ of his father's Edge Moor Iron Co. Coleman Sellers Jr. received his bachelor's and master's of science degrees from Penn in the 1870s, then became an apprentice in the machine shop at Sellers & Co. In 1886, when his father retired from the firm because of ill health, Coleman Jr. replaced him as assistant manager and chief engineer.[77]

The machine tool works was an important training ground for sons and nephews of other members of the region's engineering elite, and it was a magnet for highly skilled immigrants. Strickland L. Kneass attended the Rensselaer Institute of Technology, the nation's first engineering school, before joining the firm's injector department. His father's colleague Enoch Lewis, a Pennsylvania Railroad engineer and Franklin Institute director, sent his son Wilfred to MIT and then to the Sellers machine shop and later its drafting department.[78] William and Coleman's "warm friend" John Towne secured a similar position for his son Henry, who had studied physics at the Sorbonne. William subsequently introduced Henry to Linus Yale, with whom he went into business pioneering the manufacture of locks and cranes.[79] Sellers & Co. also drew young mechanics from Germany and Scandinavia. Carl Barth attended Norway's technical school at Horten, and upon his arrival in Bush Hill submitted examination drawings that landed him a job in the drafting department.[80]

The educational opportunities at Sellers & Co. were particularly valuable because its proprietors invested in cutting-edge research and development. William and John Jr.'s acquisition of Midvale Steel and establishment of Edge Moor Iron provided an impetus for product innovation at the machine tool works. Steelmaking and iron bridge building were in their infancy and required new tools to make production both efficient and profitable. William and Coleman supported research and development of improved gearing for power transmission, conducted mainly by Wilfred Lewis. Even Millbourne Mills on the old homestead in Upper Darby became, according to a company handbook, "a laboratory" where the "introduction of modern machinery . . . created a complete revolution in the milling practice," from "a trade" to "an art or science."[81] John Jr. and William purchased Swiss rolling machines exhibited at the centennial, installing them in a small mill owned by their cousin James on the Wissahickon Creek above Philadelphia. After two years of research and development, they made Millbourne apparently the first American mill to grind grain exclusively on rollers.[82]

Outside of the factory, the Sellers and their workforce benefited from the courses, lectures, libraries, and debates in technical and professional societies. As already noted, Sellers & Co. employees—as well as those at Edge Moor and Midvale—were active members of the Franklin Institute and American Society of Mechanical Engineers.[83] Sellers employees Carl Barth and William Thorne taught in the institute's drawing school, cultivating new pools of talent to supply the drafting offices of the region's machine makers and civil engineers.[84] In 1869, Coleman spearheaded the institute's reorganization according to a "sectional arrangement" whereby the region's leading mechanical, chemical, and later electrical engineers formed groups that resembled university departments.[85] Outside of the region, he was a corresponding member of the Société des Arts, Classe d'Industrie et de Commerce, of Geneva, and the Royal Norwegian Saint Olaf's Order. William's institutional affiliations also included the Academy of Natural Sciences in Philadelphia, the American Society of Civil

Engineers, the National Academy of Science, the Institute of Mechanical Engineers of Great Britain, and the Société d'Encouragement pour l'Industrie Nationale in Paris.[86]

While they traveled in the uppermost circles of the world's engineers, the Sellerses also promoted basic technical training for Philadelphia's less specialized industrial workers. These laborers were vital to many of the Sellerses' customers, and more broadly to the continued growth of local manufacturing. In 1873, in one of his many speeches at the Franklin Institute, Coleman called for more widespread technical education at the elementary and secondary levels. "Philadelphia stands to-day the greatest manufacturing city of this continent. In its 8339 establishments, well nigh of $200,000,000 worth of machinery is adding its products to the commerce of the country." Yet its recent growth rate could not be sustained under current conditions. "It is impossible for all the boys who desire to enter the workshops in this city to find employment" through firm-based apprenticeship programs. He concluded, "Any system of education that to mental culture adds manual skill and impresses the student with the fact that there is dignity in manual labor, and that implants in him habits of industry and frugality, is to be encouraged."[87]

Manufacturers in Philadelphia and other cities endorsed technical schooling as a replacement, and sometimes an adjunct, for the older method of craft training by apprenticeship. As female employment in the textile and garment industries expanded in the second half of the nineteenth century, the Franklin Institute gave birth to the Philadelphia School of Design for Women.[88] In the 1880s, textile makers founded the Philadelphia Textile School. Machine builders took control of the Spring Garden Institute, which the Baldwin Locomotive Works, located just across the street, effectively substituted for its internal apprenticeship system. Although Sellers & Co., Bement, Cramp's shipyard, I. P. Morris, and other metalworkers continued to operate their own apprenticeship programs in the late nineteenth century, and sometimes into the twentieth, they also paid for their workers to attend Spring Garden. Many of them, including the Sellerses, contributed to its endowment.[89]

Technical schooling at Spring Garden and other institutions was more than a simple replacement for the knowledge and skills afforded by apprenticeship. As Philip Scranton and Andrew Dawson have argued, the manual training movement represented an attempt by workshop owners to gain greater control over their employees. Schools like Spring Garden allowed them to influence curriculum, discipline, and the selection of teachers in ways that apprenticeships run by their workers did not.[90]

In an address to the Engineers' Club in 1890, William Sellers explained that the growing scale of the workforce in many firms and sectors offered a further motive for industrial schooling. The "system of apprenticeship," he observed, "became impracticable with the vast army that required to be taught the special details of an art only."[91] The mass proportions of the region's industrial labor force, together with the proliferation of mass-production techniques in the late nineteenth century, provided an impetus for public school systems to develop vocational programs. The Spring Garden Institute served as the model for the first public vocational high school in the city, which opened in 1885, not coincidentally in the heart of Bush Hill.[92]

Although the Republican machine usurped control of the school district in the 1890s, Coleman Sellers could still point to the movement for technical education as one of the sources of "American Supremacy in Applied Mechanics." "Industrial schools in connection with our common-school system are now filling a long felt want," he wrote, "and are in the near future to impress upon parents the great advantage of mechanical training as tending to increase mental activity." He waxed triumphant about the economic and physical development that resulted from the "training of the hand, that member of the human body which makes man the ruler, making him with his armed hands the superior of the most violent brute force and enabling him to tear down the mountain of minerals to gather the metals that are to serve his fellow-men."[93]

In Coleman's view, however, "Technical schools are not to be expected to make mechanical engineers." Instead, "They are expected to teach habits of correct thought, habits of correct method, to impress on the mind fundamental laws of mechanics, to teach correct business methods and habits of observation." Even higher education would not do away with the need for training in the factory. "Those who expect the Stevens Institute and the Boston Institute of Technology [MIT], or the technical schools of the universities, to give instruction that will do away with the need of actual shop practice and experience gained afterward, will be disappointed."[94] That said, higher education in engineering became increasingly important, especially for firms like Sellers & Co.

As suggested by the careers of the company's draftsmen and assistant managers, university research and training became a cornerstone of late nineteenth-century Philadelphia's leadership in manufacturing and allied science. "I am safe in saying that no profession requires a broader education than that of the mechanical engineer," Coleman maintained in his 1886 presidential address to the American Society of Mechanical Engineers.

"He must be a physicist, a merchant, a lawyer, a chemist, and he should know how to express himself in his mother tongue and be master of the modern languages far enough to have access to the scientific publications of other countries."[95] Not to mention customers overseas.

Upon his retirement from Sellers & Co. that year, Coleman accepted appointments as professor of mechanics at both the Franklin Institute and the Stevens Institute of Technology in Hoboken, New Jersey. He dedicated much of his later life to higher education. He served as the founding president of the Pennsylvania Museum and School of Industrial Art and helped Stevens president and William's former secretary, Henry Morton, develop one of the nation's premier technical colleges.[96] However, it was William and fellow trustees at the University of Pennsylvania who had the more profound impact upon higher education in the Philadelphia region.

Penn, which Coleman referred to as "a mechanism, the object of which is to shape, to mold, minds into usefulness," was the chief institution through which capitalists shaped the engineering, business, legal, and scientific professions in the region.[97] "It used to be jocularly said," claimed Saturday Club member and Penn professor Horace Furness, "that half the affairs of The University were transacted at The Saturday Club."[98] It was presumably through his contacts in that organization that William was elected to Penn's board in 1868. There, he helped build one of the world's premier research universities, heading the board's committees on scientific and engineering education and helping to oversee the medical, biological, and business programs until his death in 1905.

One of William's first projects at Penn, in collaboration with fellow board members Frederick Fraley, J. V. Merrick, John Welsh, and shafting manufacturer John Cresson, was the development of a new campus in West Philadelphia. Chosen to guide the development of Penn's campus for his "large experience in the erection of public buildings" with Edge Moor Iron, Sellers oversaw Franklin Institute–trained architect Thomas Richards's plans for and erection of the first new building, College Hall.[99] In 1871, Sellers & Co. donated $10,000 to support this and other construction.

William and his colleagues quickly developed a campus to support diverse scientific research, some applied to industry, some promoting public health and welfare. The adjacent city hospital and almshouse offered clinical experience for medical students. In 1872, he worked with Merrick, Welsh, and faculty to design a new university hospital and medical classrooms. Just six years later, they decided that new teaching methods made "a great increase in the laboratory accommodation of the Medical Department . . . absolutely necessary," and so constructed a new building.[100]

They soon added biology and chemistry labs and an observatory. In the 1890s, building on the mission of the Social Science Association, they negotiated with the city's Board of Health to accommodate a municipal bacteriological laboratory in the university's laboratory of hygiene.

Although several prominent engineers sat on the board through the late nineteenth century, William held sway over matters of engineering education. In March 1875, he proposed and the board approved graduate programs in science and new professorships in dynamical (mechanical) and civil engineering. They joined existing chairs in mineralogy, mining, and metallurgy. That summer, his friend and fellow trustee John Towne died, leaving part of his substantial estate to pay the salaries of Penn's Science Department. The trustees resolved "that in endowing permanently the Department . . . with a sum of money larger (it is believed) than has ever been given by any one person to support the teaching of applied science," the late trustee would be honored by having the department renamed the Towne Scientific School.[101]

Students at Towne received an education still rooted in hands-on mechanics. The school's course of instruction increased from four to five years in 1880, when it began offering a master of science degree in engineering. In 1884, the university contracted with the Spring Garden Institute for Towne School students to "be taught in the use of metal and wood working tools during the third and fourth year."[102] Penn replaced this program in 1890, when Sellers and Merrick oversaw the erection of an engineering laboratory and adjacent campus power plant that they "arranged so that it might form a prominent means of educating the students in Engineering, both Mechanical and Electrical." It housed a variety of boilers made by Edge Moor Iron and machinery "so diversified in type, that comparisons and tests could be conducted on a large scale."[103] The plant's smokestack joined the clock tower and turret of the adjacent classroom and library buildings in a campus that merged the industrial and the classical in its architecture as well as its curriculum.

William and other engineers linked research and teaching at the Towne School to the work of their factories and consulting practices.[104] Professor Edgar Marburg, a former employee of Edge Moor Iron, built on the firm's work in the standardization of building systems through committees he led at Penn and the Franklin Institute to examine standards for iron, steel, and reinforced concrete. In 1898, Marburg cofounded the American section of the International Association for Testing Materials, later the American Society for Testing Materials, which still operates today as an arbiter of industrial standards.[105] Two years later, William and the trustees

Figure 14. University of Pennsylvania power plant, c. 1900, with the engineering school at the left and the university library at the right. William H. Rau, *Mechanical (Engineering) Laboratory and Central Heat and Power Plant*, c. 1900. University of Pennsylvania Archives, ID: 20031017007.

accommodated a request from the American Foundrymen's Association for "a college course fitting for their business."[106]

The steam injector presents another example of educational institutions' roles in the Sellerses' networks of institutions and firms. William and Coleman initiated extensive tests simultaneously comparing different makes and models of injectors at, or in association with, the Towne School, the Stevens Institute, the Franklin Institute, the American Railway Master Mechanics' Association, the U.S. Navy, railroad companies and trade journals, as well as the Sellers & Co. research laboratory run by Strickland L. Kneass.[107] For students at Stevens and Penn, these experiments afforded hands-on experience in product testing. The navy, as well as railroad and machinery companies using large numbers of injectors, gained a detailed understanding of the strengths and weaknesses of different models.

For Sellers & Co., the tests demonstrated the advantages of their injectors vis-à-vis those of their competitors, with the endorsement of reputable third parties, now including graduate programs in engineering. Such testing also enabled the firm to offer "guarantees of performance" to customers. In 1876, William urged the Centennial Exhibition judges "to undertake elaborate tests with injectors," though the fair lacked the facilities or funds to do so.[108] At the Franklin Institute's International Electrical Exhibition in 1884, this idea served as the basis for detailed experiments on electrical apparatus in the exhibition hall itself.

Outside of engineering, Penn's role in the regional economy expanded through an 1881 bequest from iron magnate Joseph Wharton to establish the nation's first university business school. The donor worked up a plan in consultation with trustees and fellow Saturday Club members William Sellers, Frederick Fraley, J. B. Lippincott, and William's close friend Penn engineering professor Fairman Rogers, who composed the committee that oversaw the school's development. Wharton made clear that his bequest came with certain conditions in keeping with the party line of the Pennsylvania Industrial League and the Iron and Steel Association. "I should object to any important alteration of the general scheme," he wrote the board, "especially should I object to such lowering of tone in regard to national self-protection by means of tariff laws. . . . My conviction of the necessity of bold inculcation of this principle is so strong, that I at first drew up a special provision by which the endowment should revert in case of failure to uphold it."[109] After deliberation with his friend and colleague William, however, he came to understand that the language of the contract conveying his gift would bind the school to this policy.

In addition to courses in finance, accounting, and mercantile law, the Wharton School's curriculum included instruction in the basics of elite capitalists' vision of economic development and their own place in the economy and society. Its faculty would teach "how industries advance in excellence, or decline, and shift from place to place," and "how by craft in commerce one nation may take the substance of a rival and maintain for itself virtual monopoly of the most profitable and civilizing industries." Students would learn "the proper division of the fruits of organized labor between capitalist, leader, and workman," and "the nature and prevention of 'strikes.'" Taking a page out of the region's iron and steel lobby, the curriculum would cover "how by suitable tariff legislation a nation may thwart such designs, may keep its productive industry active, cheapen the cost of commodities, and oblige foreigners to sell to it at low prices while contributing largely toward defraying the expenses of its government." Rounding out this vision of the public good flowing from a well-ordered industrial economy firmly controlled by capital, graduates would appreciate "modern taxation as distinguished from the plunder, tribute, or personal service which it for the most part replaces."[110]

Following from this mission, the Wharton School and its faculty became the backbone of the Philadelphia Social Science Association. The school established an assistantship in political economy to support research on ballot reform. Other research intended to prevent strikes among industrial workers. In 1896 the school raised funds to appoint W. E. B. Du Bois "as Investigator of the Social Conditions of the Colored Race in this city," leading to his landmark study *The Philadelphia Negro*.[111] Du Bois's finding that systematic exclusion of African Americans from factory work lay at the heart of their poverty did not sit well with his patrons, who declined to renew his appointment.[112]

While the Towne and Wharton Schools occupied themselves with the mechanical, commercial, and social organization of the economy, in 1879 Coleman Sellers, as president of the Pennsylvania Society for the Protection of Animals, mounted a campaign for another new professional school. This one would serve the needs of the region's still vital agricultural and processing sectors. His friend John Welsh, then U.S. ambassador to England, had written back to Philadelphia with news of an urgent threat to American commerce. "The diseases of animals are becoming of great interest to the public" in Great Britain, he reported. "At this time there seems to be a great probability that the live-cattle trade with the United States will be suspended," owing to pleuro-pneumonia in thirteen oxen shipped to England.[113] Outbreaks like this threatened an increasingly industrial and global food system and the economic activity it generated.

In an appeal circulated among the region's business and civic leaders, Coleman detailed the imperative of a veterinary school in Pennsylvania. He estimated "that cows and the land to support them represent a money value of" $1.3 billion ($29 billion in 2010). Butter and cheese production amounted to "only one fifth less than the corn-crop of the land," and then there were meat and milk. Butter and cheese exports, he noted, generate $1 million in ocean shipping and $5 million in rail freight business each year.[114] Like other realms of the economy, he argued, the animal products trade required scientific research and education to ensure its safety and profitability.

Echoing professors at the Wharton School, Coleman's plea for a veterinary school included a call for active government involvement in the administration and protection of the agricultural sector. "We have governmental entomologists studying the habits of insects on the field, and pointing out the best methods of destroying some of the most hurtful ones," he observed. "Our rivers were becoming depleted of fish . . . and now we understand how, and do actually bring back to our streams in even greater abundance than formerly all those fish valuable for food." But the state had no such role in veterinary medicine. "An industry so vital, on which so much of the wealth of our country depends, is left without government recognition, without schools, without appreciation."[115]

Three years later, in 1882, the SPCA and Penn trustees had raised sufficient funds to start a Veterinary School in collaboration with state agricultural authorities. Together with the medical, business, and engineering schools, it made the university a leader in research and professional training. This bolstered not only the region's economy but also its competitive position in relation to other educational and industrial centers.

Post–Civil War Philadelphia's educational-industrial complex enabled the Sellerses and their allies to exert a broad influence on the region's economy and society. They determined the ways in which new generations of professionals were trained, and the standards and boundaries of the emerging professions themselves. They gained some victories in government reform, though undoubtedly the Republican machine limited their power. The civic institutions they led afforded some spaces in which they could influence economic and urban development. Certainly, the Sellerses and fellow elite industrialists created a lasting institutional infrastructure that for a time, especially in the two decades after the Civil War, supported their aspirations for Philadelphia to be a leading manufacturing center on the world stage. The educational institutions they shaped, in particular, would remain critical parts of the regional economy.

They also defined and to a considerable extent advanced an agenda that elite urban reformers across America would build upon in later decades. In more than one sense, the Sellerses and their colleagues were early Progressives, promoting an activist alliance between industry and the state, and laying the intellectual and institutional foundations of the urban disciplines, professions, and reform movements—and the relationships between them. Their economic development ambitions lay at the heart of this larger project, manifesting themselves in pursuits as diverse as the high-tariff lobby, engineering and legal research at Penn, and advocacy for civil service reform. Their networks of firms and institutions applied science and social science to technological, business, and related policy development. This pattern would characterize the "science" of economic development as it developed across the twentieth century.

CHAPTER 5

✕

EMPIRES OF STEEL

Visitors to the 1876 Centennial Exhibition witnessed not only consumer goods and the machines that produced them. They also encountered exhibits of the key tools in the Western powers' development of a global marketplace—cannons, shells, and torpedoes made of what manufacturers called steel.[1] Most of the weapons at the exhibition came from Germany and England. But one Philadelphia exhibitor distinguished itself for its steel railway tires, axles, and super-hard tools for cutting metal, apparently the first such items made by the Siemens method of steel production. This established William Sellers's Midvale Steel Co. as the leading scientific steelmaker in America.[2]

Although it did not display machines of war at the fair, a year earlier the firm had taken its first order from the U.S. Navy's Bureau of Ordnance. This was the first time the navy bought steel gun forgings from an American manufacturer. It initiated Midvale's long history as one of the foremost contractors in what would soon become a massive military-industrial complex centered on the production of steel. The company's cannons, projectiles, and armor plate made possible the warships that would aid American capitalists in penetrating overseas markets.

Looking back in the first article in the January 1900 edition of the *Journal of the Franklin Institute*, Coleman Sellers cited military contracting as a prime driver of American and European industry in the late nineteenth century. "The nations have been increasing the power of their fighting ships year by year, adding new improvements in machinery, new means of resisting attack from other ships, new guns and newer projectiles to pierce the new armor plate." In an apologetic gesture, he expressed hope that these ever more lethal technologies would lead to "decreasing opportunity to bring the results to actual test in battle." He cited evidence from the recent Spanish-American War, which "found our trained marine force better equipped for the operation of these fighting machines, and in the far East and in the West Indies our guns swept a maritime power out of existence with upon our part scarce the loss of a man."[3]

In the decades following the centennial, the United States became a global military and commercial power. The federal government and allied capitalists built a network of naval bases and coaling stations, the Panama Canal, and a steel navy to escalate their influence in Latin American, Caribbean, and Pacific markets.[4] While New York was the financial capital of this new empire, Philadelphia was its workshop. Midvale and affiliated firms, including Joseph Wharton's Bethlehem Steel and Cramp's shipyard, dominated military contracting. The military-industrial complex thus became an increasingly important engine of economic development for cities as well as nations.

The story of Midvale, which William acquired with fellow Saturday Club member Clarence Clark, shows how industrialists shaped new forms of corporate capitalism that in turn built a new national and international capitalism. Although their networks of elite businessmen proved unable to defeat the political machine in Philadelphia, their federal and overseas projects gave men like William greater influence at larger scales. His interest in steel brought him in contact with the foremost manufacturers of England, Germany, and France, and with customers in Latin America and other industrializing regions of the world. Metallurgical research and management reform at Midvale furthered his rationalization mission and helped extend his sway over national economic policy and in international manufacturing. The firm's leadership in chemical analysis and management reforms, together with William's connections, helped make it a core supplier of the railroads and the U.S. Navy.

William and John Jr.'s three big companies, Sellers & Co., Midvale, and Edge Moor Iron, enjoyed international markets and often played complementary roles in economic development.[5] (Millbourne Mills, which John Jr. still ran, remained a more modest concern.) Sellers & Co. and Edge Moor served the specialized needs of shipbuilders, civil engineers, and railroads, making custom tools and bridges. Midvale supplied steel armor and rails to some of the same customers. Edge Moor's engineers designed standardized construction systems for steel buildings in booming cities.[6] At Midvale,

William sponsored Frederick Winslow Taylor's earliest experiments in scientific management. In these and other ways, they helped fashion the technologies, markets, and metropolitan environments of the second industrial revolution, when cities of iron, coal, and railroads became cities of steel, electricity, and later automobiles.

Building Corporate and Scientific Power

Leveraging the capital they acquired in machine building, the Sellerses gained control of the firm that would become Midvale through a series of strategic and accidental moves, aided by failures in production. Their ties to powerful railroads and financiers also proved critical in developing the firm and its early business. Mirroring their approach at their tool works, they used the company to transform steel manufacturing from a rather haphazard enterprise to a laboratory-centered industry. Supported by the trade associations led by William Sellers and Joseph Wharton, they shaped national policies related to the late nineteenth century's most important material. Midvale's research and development, coupled with this lobbying, helped define steel in scientific terms and delineate its place in the world economy.

William made his first investment in steel production in 1864. With Pennsylvania Railroad president J. Edgar Thomson and former Philadelphia, Wilmington & Baltimore president Samuel Felton, both members of an investing group known as the "Philadelphia Interests," he formed the Pennsylvania Steel Co. near Harrisburg. They engaged engineer William Butcher from Sheffield, England, the world's foremost center of early steel manufacture, who recruited workers from his hometown.[7]

At the end of the Civil War in 1865, William Sellers journeyed to Britain on one of many trips he and Coleman took to study machine and steel production in Sheffield, Manchester, and other cities. In October, Coleman wrote to Escol, reporting, "Wm. Sellers has just returned from England and tells much that is interesting about the steel works of the Bessemer & other processes. He says that large machinery has been much improved during the past 6 years," since his last visit.[8] These voyages, like Escol's trip back in the 1830s, sustained intimate transatlantic networks of engineers and a rapid pace of technology transfer.

William reported the results of his trip at the November meeting of the Franklin Institute. "Whilst in England, I noticed great progress in all the interests, the most remarkable being in the manufacture of steel."

Through the 1860s, steel making would remain something of "a black art and a crap shoot," according to John K. Brown.[9] Engineers experimented for years with different combinations of metals and chemicals, and with a variety of production strategies before determining reliable methods.

Yet it was clear to William that new processes to produce steel in quantity heralded big changes in metal manufacturing. "Formerly this business was confined to the manufacture of steel for cutlery purposes and other small objects, the use of it in large masses being unknown," he told his colleagues. "But within the last six years, the general introduction of the Bessemer process, as well as that employed at Mr. Krupp's works in Germany, have revolutionized the trade." The result was "that work which, under the old system, would require an immense number of hands, can now all be done . . . by a few, making the steel thus produced comparatively cheap, so that it can be applied to ordinary purposes where iron has heretofore been used."[10]

This meant a new era in metallurgy and heavy industry, one in which capital gained further power over labor, the organization of work, and the built environment. William's journey inspired much of Sellers & Co.'s innovation of hydraulic tools, since he noticed that hydraulics allowed a single worker to handle with "facility large masses of molten metal . . . ten tons often being taken off at a heat."[11] Resulting labor and cost savings would enable manufacturers, railroads, builders, and the military to replace common iron with the stronger material of steel. This would affect machines, transportation, tall buildings, and the military on a massive scale and increasingly through mass production. But first Sellers and his colleagues would endure a series of failures in production, which would influence much of Midvale's later research.

In 1866, William and his Pennsylvania Steel partners purchased rights to the Bessemer production process from its U.S. license holders, the works of Alexander Holley at Troy, New York. Holley was an American who had studied in Sheffield. They hired him to design and construct their new plant, replacing Butcher, who had been slow to get the concern off the ground.[12]

Although William was a minor partner in Pennsylvania Steel, he and John Jr. soon came to control a local steel mill that Butcher had established after Holley replaced him. Butcher's principal backer in this Philadelphia venture was Philip Syng Justice, the first merchant to import steel rails from England. Justice had determined he could turn a greater profit making steel rails and railroad car tires locally than by importing them.[13] He and Butcher purchased a property with ample groundwater (necessary for steelmaking) in the Nicetown section of the city from the German-town & Norristown Railroad, near its junction with the Reading Railroad. The latter company became the Butcher Steel Works' first customer in 1867, purchasing the first steel castings made in the United States.

Butcher struggled in Philadelphia, as he had in Harrisburg. He joined the Franklin Institute, and Asa Whitney & Sons tested his cast steel car wheels with favorable results.[14] However, the R. G. Dun credit agency reported that the Butcher Works "has hazards & risks attending" its financial position. Butcher had "failed it is alleged for 50,000 (Sterling)" back in Sheffield, and his "means are [considered] small." Compounding these issues, Dun discerned, his partner Justice "is looked upon by Careful & experienced men as getting much extended," with $30,000 in cash debt and a mortgage on their plant held by the Reading Railroad.[15] When the works contracted with Sellers & Co. for a seven-ton steam hammer and other machine tools in 1868, Butcher and Justice lacked the cash to pay up front. They therefore transferred a portion of the company's stock to William and John Jr., giving them a significant, though not yet a majority, interest in the firm.[16]

The following year, the Butcher Works began making steel in open-hearth furnaces according to the Siemens-Martin process, but the challenges of developing this new method brought considerable financial losses. Justice turned to investment bankers Edward and Clarence Clark, brewer Samuel Huston, and William and John Sellers Jr. for an infusion of additional capital. In June 1869, R. G. Dun reported, "The original stock will be sunk. W. Sellers & Co, E. W. Clark & Co. & other very wealthy men," including William's Bush Hill neighbor Asa Whitney, "control it & hold $200,000 stock & under the charter are liable for the [amount] of their stock, which makes the [company] very responsible for contracts."[17] The new owners reorganized the company, with Huston as president and Butcher as manager.

In the fall of 1870, the Butcher Works contracted to furnish twenty-four hundred tons of steel to the Keystone Bridge Co. of Pittsburgh for what later became known as the Eads Bridge over the Mississippi River at St. Louis. This was the first steel bridge built in North America, a record-breaking, double-deck span with a road running above railway lines. Long tied to Philadelphia capital, the St. Louis promoters of this project allied with Pennsylvania Railroad officers J. Edgar Thomson, Tom Scott, and their "Philadelphia Interests." E. W. Clark & Co. secured a loan of $250,000 for James Eads's bridge company, the builder, on the condition that Thomson and Scott—the majority owners of Keystone and close allies of the Sellerses and Clarks—instruct their manager, the "little white-haired devil" Andrew Carnegie, to subcontract for steel with the Nicetown firm.[18]

This contract would prove the downfall of William Butcher. Carnegie told Eads that William Sellers and Samuel Huston already had "no more regard for [Butcher's] opinion, or his promises than if he were totally unknown to them."[19] One later employee of the works characterized Butcher's "rule of thumb" in production and management as "reckless and utterly wanting in system."[20] By contrast, Eads was a serious engineer whose "rigid tests and demands for exceptional workmanship," according to historian Geoffrey Tweedale, "were to the limits of available metallurgical knowledge (and perhaps a little beyond)."[21] This experience likely influenced William Sellers's later investment in metallurgical research.[22]

Using the Bessemer and Siemens processes, Butcher failed repeatedly to make suitable anchor bolts and staves, the main structural steel parts of the bridge, even as the firm's owners raised its capital to $1 million, took out "heavy mortgages for money advanced," and invested over $500,000 in new furnaces, machinery, and buildings.[23] Eads then paid the Chrome Steel Co. of New York a $15,000 royalty for a license and training for Butcher to produce Chrome's patented steel. When Butcher botched this attempt as well, Keystone finally canceled his contract. Butcher had filled much of the Eads orders, but not enough and far too slowly. His Philadelphia backers terminated their association with him in the fall of 1871. Understandably "quite dispirited by his lack of success," Butcher leased another steel mill near Lewistown, Pennsylvania, where forty of his English workers reportedly joined him. He died soon thereafter.[24]

In the aftermath of Butcher's departure, the Sellerses, Clarks, and other stockholders grappled with the management, production, and financial conditions of the company. The Eads job was not the only costly disaster. A contract to furnish the Reading Railroad with two thousand tons of rails in 1871 resulted in little more than scrap metal and further damage to the firm's reputation. Butcher had attempted to use old rolls in the Reading order, adding too much phosphorous to the steel, which made the rails overly brittle and caused them to break when unloaded.[25] In response to such errors, William hired Yale University–trained metallurgical chemist Charles Brinley in 1872. William changed the firm's name to the Midvale Steel Works, ridding it of what R. G. Dun termed "the 'incubus' of even the name of Wm. Butcher."[26]

The company's substantial debt led to protracted struggles among its owners. In early 1873, they issued additional stock and mortgage bonds, though the firm remained on the verge of bankruptcy. Huston's health deteriorated, and on May 13 he signed an agreement with the Sellerses and Clarks that relieved him of all liability for Midvale's debts in return for repayment of his investment from a guaranty fund they established.[27] This gave the Clarks and Sellerses all but 350 of the company's 5,000 shares of stock. William became its president and hired a new superintendent. But the situation worsened further before it got better. Later that month, the Dun agency reported, "There is at present [considerable] depression of confidence in the minds of initiated parties as to the responsibility of this concern."[28] E. W. Clark & Co. went bankrupt in the financial panic that fall, brought down by the failure of their old associate Jay Cooke. William resigned from the Centennial Finance Committee to attend to Midvale, where only about seventy-five men remained employed. He traveled to England in an unsuccessful attempt to find a buyer for the firm.[29]

In 1876, not having received his money despite a turnaround in Midvale's business, Huston took his former partners to court. Although R. G. Dun speculated in August of that year that "it is thought it will be amicably settled," this case dragged on until 1880, when Clarence Clark brought another suit against the company.[30] Clark had not been paid the interest on mortgage bonds he held as collateral for loans from his father, Edward, who had since recovered from the financial panic. However, his legal action may have been part of what historians Charles Wrege and Ronald Greenwood have argued was a scheme on the part of the Clarks and Sellerses to dupe Huston out of his share—rightful or not—of the company that William Sellers had since resuscitated.[31] In 1878, the Sellerses and Clarks floated another $500,000 in bonds, and the following year Dun claimed that "they are very busy, are extending the capacity, making extensive improvements to the [property] & appear to have very fair prospects."[32] Whatever the intentions of Clarence Clark's claim or the merits of Huston's, these legal battles resulted in a mortgage foreclosure. The Nicetown property was sold at a court-ordered public auction in October 1880.

The man who purchased the land, buildings, and machinery for $450,000 was William Sellers. He raised the capital with the help of his brother John Jr., the Clarks, Pennsylvania Railroad director James Wright, and Sellers & Co. partner and business manager James Brooks. With the title to the property cleared, these partners reconstituted the works as the Midvale Steel Co.[33]

While he was busy sorting out Midvale's finances and ownership, William put twenty-five-year-old Charles Brinley to work solving issues

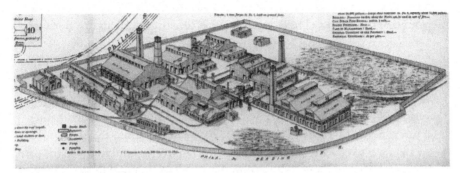

Figure 15. A drawing of the Midvale Steel works in 1879, showing its large plant connected by rail lines. *Hexamer General Surveys*, vol. 14 (Philadelphia, 1879), *plates 1332–1333.* Map Collection, Free Library of Philadelphia.

of metallurgy and production. The Eads and Reading jobs had left upward of two thousand tons of scrap in the yard. However, after systematic chemical experiments, Brinley was able to turn the metal into durable steel rails that the firm sold to the Reading line. He also succeeded in making the first three thousand open-hearth steel railway car tires in the United States, all of which held up in service. At the same time, the Pennsylvania Railroad submitted trial orders for steel axles from several hearths in the East. The Pennsylvania deemed Midvale's product superior and thereafter awarded the company large and frequent contracts. Brinley's track record in production quickly improved the firm's reputation, and it heralded what Tweedale has termed "a new era . . . which demanded that steelmakers be led by books and a laboratory rather than by the older practiced techniques."[34]

In 1874, Brinley's old roommate at Yale, Russell Wheeler Davenport, came on board. After graduation in 1871, Davenport had taught for a year at Yale's Sheffield Scientific School and then pursued postgraduate studies at the Royal School of Mines in Berlin. Before returning to America, he "visited all the most important iron and steel works in England, Wales, France, Austria, Germany, and Belgium."[35] Soon after Davenport arrived in Nicetown, the plant's superintendent died of malaria, and Sellers promoted Brinley to his post.

Brinley and Davenport lived their work. By Brinley's account, "Davenport found his friend living in a workman's house of four rooms. The two on the ground floor had been converted into a laboratory. In one of the upper rooms there was a large water-tank and some lumber." The fourth room "was scarcely large enough for a single bed, a washstand, a chest of drawers, and a couple of chairs." Davenport squeezed in. According to Brinley, "not a night passed without a visit to the furnaces, followed by a round of the Works. There was not much time for social affairs, nor was there a great sense of loss on that account."[36]

Nicetown was a kind of industrial suburb-in-the-city. Up the hill in the railroad suburb enclave of Southwest Germantown lived many of the region's elite. Down in the valley between Germantown and the worker housing of Nicetown, next to a malarial swamp, Midvale and other manufacturers began to locate in the 1860s. "Between the house and the Works ran a branch of the Reading Railroad, over which day and night trains of coal cars rattled by on their way from the Falls of Schuylkill to Port Richmond on the Delaware," wrote Brinley. "A woman of the neighborhood came in to cook; and meals were served in the lower front room, which was nearly filled by a balance and stool, a machine for physical tests, a table, and a broken-backed armchair cushioned with an old coat that had belonged to a former chemist who had died of malaria."[37]

Brinley and Davenport took charge of laboratory research from their kitchen and living room and supervised production at the plant. This all was "subject, of course, to the orders of the President of the Company, Mr. William Sellers, who," Brinley added, "very rarely visited the Works, although [he was] in constant touch with his lieutenants."[38] Their living arrangement lasted until 1877, when Brinley married and moved to Germantown.

As a young, academically trained, and somewhat aloof manager who had not come from the ranks of the workshop, Brinley did not make fast friends among Midvale's mechanics and puddlers. With a silk handkerchief in the breast pocket of his coat and an aroma of perfume about him, he wore his social class openly.[39] His disdain for the working class mirrored but probably exceeded that of his employer. Yet, with a mandate from William to introduce systematic, scientific methods, Brinley, again by his own account, prevailed against the old ways of his workforce. "The practical men that remained in the Works objected to methods they did not understand, and they objected to being taught, as they said, by boys; some threw up their jobs; others submitted, but with grim predictions as to results."[40]

William's investment in metallurgical research and production reforms enabled the firm to shift from the competitive business of making rails and axles to manufacturing high-quality, high-priced, batch steel, including for the military. Thanks to changes instituted by Brinley, "The results were the successive placing upon the market by Midvale of various high grades of open-hearth steel that before had only been produced by the crucible process."[41] By the early 1880s, the workforce reached six hundred men.[42]

As at Sellers & Co., management gained firm control of the plant thanks partly to the rationalization of production. In 1876, one of Brinley's many memos to his shop floor supervisor reveals some of the ways he tightened supervision, reporting, and production standards. "See that every ingot beaked bloom or punched bloom or rolled tire is reported to the office as scrap," he wrote. At the tire mill, "let me know at once if any thing is going wrong, as bad ingots, bad blooms, or bad rolling. Think for yourself as to how you can best systematize every thing in connection with the tires, and suggest anything that occurs that will prevent errors, save labor and hasten delivery." In success and failure alike, the work must be meticulously recorded. "Make a daily report . . . every afternoon, even if the same report goes in more than once. . . . Read these instructions over from time to time and devise a regular routine for your days work in accordance with them."[43]

Like the drafting room at Sellers & Co., Brinley's notebooks were the central repositories of systematic data collection at Midvale. He copiously documented everything from chemical tests to the construction of new hearths, the performance of machinery, and the division of labor. He carefully tracked the progress of each order received from locomotive and car builders in the Northeast and Midwest. He outlined in detail the duties of the heads of departments at the firm. These consisted of a chief engineer's office; metallurgical department; tire and rolling mills; the hammer, machine, carpentry, and blacksmith shops; a bricks and bricklayers department that built and maintained the furnaces; and rigging and yard outfits responsible for moving and storing materials. To ensure efficient communication between these departments and the main office, he instituted a color-coded system of orders and internal memos.[44] Brinley and his department heads paid close attention to each function performed in the production process, working to maximize its efficiency.

While compiling statistics for the 1880 census, Brinley noted several cases in which personnel realignments enhanced productivity. Prior to 1875, the tire mill had rolled an average of sixteen and a maximum of eighteen tires in ten hours. The roller "was replaced and then changes made in the rolling gang, and an agreement made with each man that he should be paid so much per tire rolled according to the weight." Under this payment system, "production has grown to 28 to 40 tires in 10 hours according to size."[45] Through piecework, and showing what could happen to workers who resisted production reforms, between 1875 and

1880 Brinley increased the output of Midvale's furnaces and hammers an average of about 110 percent while lowering labor costs from 2.28 to 1.26 cents per pound of steel produced. Total output in these years climbed from 2.6 to 14.6 million pounds, and wages paid rose from $59,472 to $184,099.[46] The new system grew management's control over labor by effectively buying them off.

One of Brinley's chief collaborators was Frederick Winslow Taylor, who arrived in 1878 after an apprenticeship at a small pump-making shop on the edge of Bush Hill. From an elite Germantown family, he was Clarence Clark's brother-in-law. William Sellers helped Taylor gain admission as a corresponding student at the Stevens Institute of Technology in 1881. He spent the next two years studying metallurgy, ores, and fuels in the chemical department of the Towne School, all the while working at Midvale. In 1881, Taylor and Clark found time to win the U.S. Open doubles tennis championship.[47]

Midvale Steel is best known among historians as the birthplace of scientific management, the ultra-rationalization of labor organization and production.[48] Taylor later presented scientific management (or Taylorism) as his own personal crusade and triumph over shop floor workers set in their ways, and he certainly did popularize it.[49] But in fact the reforms he claimed to originate fit squarely within the research and development program initiated by his employers in Nicetown.[50]

Management challenges at Midvale were mounting when Taylor was hired. The firm had recently increased its workforce to four hundred. Then, in 1879, William took on a job to roll the structural steel for the Brooklyn Bridge, a time-sensitive project as the cables for the bridge were already in place, awaiting the steel that would hang from them. As Brinley related, Midvale brought in "nearly a hundred strange men—gathered from anywhere." The plant had an old rolling mill dormant since Butcher's departure, though "No one at the Works had any knowledge of rolling-mill practice."[51] The workers balked at the pace of work Sellers and Brinley demanded. They struck, and management replaced them.

The challenges of organizing their large, fluctuating workforce and time-sensitive contracts led Sellers, Brinley, and Davenport to pursue research on the productivity of machine tools and to refine the piece rate system already in operation at Midvale.[52] Much of this work fell to Taylor, whom Brinley promoted to gang boss in 1879, shop superintendent in 1881, and master mechanic soon thereafter. After Davenport assumed the superintendent's post, in 1884 Taylor became chief engineer.

In 1880, William Sellers put Taylor to work on experiments comparing the performance of different grades and shapes of steel cutting tools.
These tests built on earlier research at his machine tool works, and they
helped establish accepted best practices.[53] Similar experiments were undertaken at the same time in Philadelphia's principal rival in machine tool
and steel manufacturing, Manchester, England, at the local Association
of Engineers, School of Technology, and Whitworth & Co. This was an
important part of the competition between engineers in the two cities to
dominate the world's steel sector.[54]

Taylor's most celebrated experiments, however, came in the area of
labor management. Armed with a stopwatch and a clipboard, in 1881 he
began an intensive study of the time it should take a worker to perform
each motion or task in the factory. In 1883, William transferred Wilfred
Lewis from Sellers & Co. to join Taylor at Midvale, expanding that firm's
capacity for research and development. Their work helped expand production, building on earlier work by Brinley, Lewis, and others at William's
various firms. In an article on shop management at the three Philadelphia
shops with established piece-rate systems, a correspondent for the *Engineering Magazine* claimed, "The general effect of piece-rates at the Baldwin,
Midvale, and Sellers establishments has been to double, and in some cases
[at Midvale] treble, the output of the plant, increase the yearly earnings of
the men, and greatly reduce the piece-cost to the management."[55]

"It is almost needless to say," the correspondent continued, "that the
Midvale men do not join trade societies founded on 'equal rights and equal
pay for all,' because the immediate result would be a lowering of their own
pay by their own choice, which is a conclusion most unlikely to be reached
by the workman."[56] Taylor's refinement of the piece-rate system not only
helped the firm refine its labor management to meet the demands of intensive batch production. It also effectively solved, at least for Midvale's
managers, "the labor problem" that increasingly faced large manufacturers
in the late nineteenth century.

Midvale operated differently from Sellers & Co. and Edge Moor, with
a different logic of labor and management, yet all three contributed to
the evolution of production in the second industrial revolution. Midvale's
huge orders for uniform car tires, axles, and rails brought incentives to
develop systems resembling mass production, even if the firm did not
serve the mass consumer markets that characterized later automobile or
radio manufacturing. Sellers & Co. remained mainly a custom and small
batch producer, while management and production at Edge Moor Iron fell

somewhere between Midvale and the tool works. Edge Moor relied on bulk (or nearly mass) production techniques for its structural steel bridge and skyscraper parts. But it combined them in a nearly infinite variety of configurations depending upon where and what it was building. Both Edge Moor and Midvale depended upon the massive, custom tools of Sellers & Co. to make bulk or mass production possible, and Edge Moor designed and furnished ironwork to meet the particular needs of a Sellers & Co. crane installed at Midvale in 1879.[57]

Beyond Midvale's influential owners and shop floor reforms, the foundations of its leadership in high-end steel lay in the constant tests undertaken by the two Yale chemists and their colleagues in related firms and institutions. They conducted experiments, for example, "to determine the effect of pickling hammered steel in sulphuric acid on its physical properties," "to investigate the influence of working & finishing a high carbon steel . . . at different temperatures upon its specific gravity and cutting tool properties," and "to determine the influence of a radial crystalline structure in a crucible tool ingot."[58] Using different sources of iron and a myriad of heating and treatment techniques, they worked to develop standards of uniformity in the chemical composition of manufactured steel.

In addition to their own tests, Midvale's managers were active in networks of metallurgists and manufacturers defining the properties of the new material. They sent their tires and axles for testing at Asa Whitney & Sons in Bush Hill and the Cambria Steel plant in Johnstown, Pennsylvania. They corresponded with the superintendents of other furnaces in the Middle Atlantic and New England and with professors at Penn and the School of Mines in Stockholm. They collected French, English, and American publications on steel.[59]

Brinley and Davenport's experiments enabled William Sellers to become a leading representative of the steel industry in America and abroad. Like his "lieutenants" at Midvale, he remained active in the transatlantic transfer of steel technologies. He became involved in the Iron and Steel Institute of Great Britain, and kept a close eye on European production. In 1878 he joined with Alexander Holley in purchasing half of the right to the patent casting process of the Compagnie des Fonderies et Forges de Terre-Noire, the second-largest steelmaker in France.[60]

William's investment in Midvale and its various trials cemented his devotion to the high tariff. Thanks to an eighteen-month lobbying campaign William spearheaded through the American Industrial League and the Iron and Steel Association, the U.S. secretary of the treasury agreed to reclassify the imported Siemens–Martin metal as steel, after it initially broke into the American market under the lower iron tariff, a status that undercut domestic steelmakers.[61] This controversy inspired attempts by American and European metallurgists, including those at Midvale, to formulate a standardized classification of iron and steel. According to historian Thomas Misa, however, "differences in industrial practices as well as linguistic problems paralyzed the effort."[62]

Midvale Steel allowed William to shape new, in some cases seminal, business and production practices of the second industrial revolution. The alliances of capitalists responsible for steel firms' establishment and their big contracts constituted especially important networks in the formation of national capitalism. Brinley and Davenport's chemical experiments and their reform of production and management extended the Sellerses' rationalization mission beyond the machine shop and into the foundry and the science of metallurgy. Through Taylor's later promotion, they would help shape the new science of corporate management. However, it was Midvale's military contracting that made the deepest contributions to positioning Philadelphia and the United States as a global industrial power.

Fortress Philadelphia

Like his great-grandfather in the Revolution, William was a "fighting Quaker." In a letter to Joseph Wharton in the 1890s, he explained, "I am still a member" of the Society of Friends, "although I suppose I ought not to be, as I cannot concur in some of the teachings, . . . notably that of nonresistance," the Quaker compulsion to avoid war at any cost.[63] Since at least the Civil War, Sellers & Co. profited from their close association with the U.S. Army and Navy and also counted foreign militaries among their customers. At Midvale, these ties became even more important. For Philadelphia industry more broadly, the federal government's commitment to the project of building a steel navy made military contracting a significant driver of regional economic development.

In the 1880s Congress began to fund a robust military-industrial complex, replacing the nation's prior pattern of buying arms almost only during wartime. The navy received large appropriations to build modern steel warships. This policy in turn motivated the navy to form a powerful and relatively permanent network of public and private plants for arms production and testing. Most of this buildup took place in and around Philadelphia.[64] Together, Midvale Steel, Wharton's Bethlehem Iron Works, Carnegie's steel plant in Pittsburgh, Delaware River shipbuilders Cramp

Figure 16. A forty-ton crane built "for placing the heavy armor plates on the new warships" at the navy yard at Portsmouth, Virginia, in the 1890s, one of many Sellers & Co. custom machines for the navy. William Sellers & Co., *Illustrated Catalogue and General Description of Improved Machine Tools for Working Metal* (Philadelphia: J. B. Lippincott, 1895), 317.

and Roach, and the Philadelphia Navy Yard made up what historian Thomas Heinrich has called "Fortress Pennsylvania."[65] Three of the five warships that defeated the Spanish in the port of Santiago de Cuba in 1898 were built at Cramp, their cannons made at Midvale. Sellers & Co. played a part in arming both sides of this conflict, as the firm sold shafting and machine tools to Cramp and to all the navy yards on the East Coast, as well as to the Spanish Ministry of Artillery in Havana.[66] Other military suppliers in the region during the late nineteenth century included engine builder I. P. Morris, the Disston Saw Works, DuPont's gunpowder mills, and scientific instrument makers.[67]

Philadelphia machine makers had supplied the American military since the Revolution. But following the Civil War the navy turned to English manufacturers for all its steel cannons, since U.S. steelmakers could not yet meet the technological demands of these orders reliably. The navy subsequently sought to develop U.S. firms' capacity to build steel armaments, again especially in Philadelphia.

As production improved at Midvale in the mid-1870s, the navy's Bureau of Ordnance issued its first contract for steel from a domestic producer. William's newfound control of the company and his strong relationships with navy engineers surely helped him secure this job. "The forgings were for little howitzers to be used at coast-guard stations to throw a line to ships in distress," Brinley wrote. "From this small beginning the gun industry grew steadily at Midvale, and at Midvale alone . . . with the development of a complete system of experimentation and of records, until, in 1882, forgings were made for parts of guns of six-inch caliber."[68] While the Navy continued to buy gun forgings from the leading cannon makers of England, Whitworth & Co. and Cammell & Co., Midvale became its main supplier of howitzers and breech-loading rifles, steel armor plate to protect ships, and shells to pierce the armor of enemy ships.[69]

This new product line helped make William Sellers a leader in a distinct branch of the steel industry. It may have offered Midvale a respite from competition with Carnegie, whose mills were driving down profit margins in the railroad steel business.[70] It certainly fueled the firm's research and innovation, including continued technology transfer from Europe.

The navy's technical demands inspired heightened investment in Midvale's regime of chemical and physical testing. Rails, tires, and axles needed to be sturdy, but cannons required superior levels of strength, as the steel navy arms race pitted new high-explosive shells against thicker and tougher steel armor plate protecting warships and the men, guns, and explosives on board. The company and the navy subjected the cannon and armor plate produced at the works to extensive tests in Nicetown and at several federal facilities.[71] Brinley and Davenport's chemical analysis reports helped overturn the patent rights on armor plate held by the Krupp company of Germany, releasing the U.S. Treasury from royalty payments on every pound of armor produced for the navy.[72]

When Brinley departed Midvale to manage Philadelphia's Franklin Sugar Refineries in 1882, William promoted Davenport to superintendent. He also sent him on a tour of the principal steel plants in England and to the Creusot works in France, where he studied production methods for the heaviest kinds of forging.[73] The following year, co-owner Clarence Clark, though a banker, brought back from England specifications for Sir Joseph Whitworth's patents for hydraulic machines to compress and forge steel. These proved of use to both Midvale and Sellers & Co.[74]

Figure 17. Breech-loading rifles destined for navy gunships, on the shop floor at Midvale Steel, c. 1901 (#1974362_071, Pictorial Collection, Hagley Museum and Library). Hagley Museum and Library.

Coleman went to Europe in 1884, visiting engineering schools, professional societies, and dozens of machine builders and steelmakers in Germany, Sweden, Norway, and Britain. As business partner of a leading competitor, however, he was denied a tour at the Krupp plant.[75] His visit to Whitworth's forging press in Manchester was the highlight of the trip. He wrote to Coleman Jr., "I have been scheming to see this part of Whitworth's steel works ever since I have been in England."[76] He sent regular, copious reports to William, relating the details of workshop organization, steel production, and machine tools used for working steel. William was most interested in Whitworth's techniques for making ordnance. He wrote Coleman often, inquiring about this and other topics of interest to his firms.[77]

The information collected by Davenport, Clark, and Coleman in Europe helped sustain the Sellers firms as world leaders in machine tools, bridge building, and—as U.S. military spending skyrocketed in the 1880s—the production of ordnance. In 1883, Congress funded the construction of gun foundries for the army and navy. Before these plants were built, the Philadelphia-based Gun Foundry Board visited Midvale and other military contractors of the United States, England, France, and Germany.[78] Whitworth & Co. and Midvale Steel were subsequently the only manufacturers the navy deemed capable of making its largest cannons. For smaller guns it developed networks of production in which Midvale collaborated with the Washington Navy Yard, Pittsburgh Steel Casting, Pratt & Whitney, and Samuel Colt's armory in Connecticut.[79] By the mid-1880s, Midvale was turning out such a massive volume of gun forgings that the navy sent a serviceman to work full time at the Nicetown plant inspecting its output.[80] The firm continued to dominate American ordnance production into the twentieth century.[81]

Sellers & Co. also influenced military production and technology in this era. It supplied the government's gun foundries, navy yards, and arsenals with tools, including a custom-designed "bullet machine" and "terminal velocity targets" for ordnance tests.[82] The Sellers standard system of screw threads spread beyond the United States in the 1870s in large part through the millions of dollars in contracts that fellow machinery firms Pratt & Whitney and Brown & Sharpe filled for tools and training for German and other European state armories.[83] In the early 1880s, Sellers & Co.'s correspondence on matters of cost accounting and shop floor supervision with U.S. Army captain Henry Metcalfe helped reorganize the Ordnance Department and the Frankford Arsenal.[84] Metcalfe's work on information flow in manufacturing in the mid-1880s would influence Taylor's later development of scientific management.

In the nation's capital, William's authority was such that he could dictate machine design to the Navy and War Department. As related by Alexander Outerbridge and Coleman Sellers Jr. in 1890, the navy "sent to leading manufacturers of machine tools elaborate specifications for . . . a leviathan lathe," 128 feet long, to make sixteen-inch cannons. Disapproving of the government engineers' design, William "refused to bid upon them; but he caused new designs to be drawn embodying new principles, differing radically from the Government's drawings." He "explained his design to the Board of Engineers in Washington, pointing out the merits of his plans, and so thoroughly convinced the Board of their superiority that they adopted the Sellers' plans and discarded their own." Sellers & Co. built the five-hundred-thousand-pound lathe. Once installed in the Naval Gun Factory in Washington, it "attracted the attention and admiration of engineers from all parts of the world."[85]

Scientific steel and specialized machine tools made possible the construction of a steel navy. Sellers & Co. and Midvale Steel's leadership in late nineteenth-century production was, in turn, supported by hefty military contracts that required and paid for extensive research and development. As they made weapons and armor on an ever-greater scale, manufacturers facilitated and profited from the ascendancy of the United States to global political and economic power.

The federal government and its military-industrial complex thus came to play a major role in shaping not only industrial practice but also the broader economy of Philadelphia and other manufacturing regions. By the beginning of the twentieth century, Midvale employed three thousand men, while Bethlehem and the next three largest steel plate makers in the area together added another four thousand workers engaged largely in military contracting. Cramp and the two other big private shipbuilders on the Delaware River (not including the navy yard) counted over twelve thousand employees. At the same time, Sellers & Co. employed close to eight hundred men.[86] Most of these jobs paid good wages compared with other manufacturing work, partly because many were high-skilled positions, and partly since government contracts generally offered higher profit margins than private ones.

As long as William, together with steel magnates such as Wharton and Carnegie, held influence over the military's investment and engineering policies, "fortress Pennsylvania" would prosper. In addition to Sellers's and Wharton's trade associations, the Franklin Institute weighed in on federal engineering and administrative decisions, sending regular recommendations and sometimes remonstrances to Congress. William was often directly behind this lobbying.[87] His stances on particular matters of technology and trade were surely in the interest of his firms, but they were also consistent with his understanding of Philadelphia's and the nation's interests in capturing world markets on the strength of their engineering.

The Sellerses' position at Midvale, however, shifted radically in 1886. In the face of yet another suit brought by Huston, James Wright and Edward Clark sold their majority share of the company's stock to Charles Harrah Sr. and Jr. The Harrahs had made their money building railroads for the Brazilian government, part of Americans' large involvement in South American industrialization. Charles Sr. replaced William as president. The Sellers brothers continued to hold Midvale stock and sat on its board of directors into the twentieth century. But they did not control its management and became embroiled in almost perpetual quarrels and litigation with the Harrahs over nonpayment of dividends. Although the

Harrahs' management was competent, they did not encourage the level of experimentation that William had maintained at Midvale. This alienated researchers like Taylor. Relations between the Nicetown firm and Sellers & Co. soured to such an extent that Taylor was no longer welcome to simply take the train down to Bush Hill to consult with colleagues.[88] In most senses, the Sellerses exited Midvale.

The legacy of Midvale Steel, however, looms large over the history of industrial development. It is mostly a story told by Taylor, principally about himself. His disciples would adapt Taylorism to settings as diverse as machine building plants, department stores, and government bureaucracies.[89] However, this obscures how his work was a product of the problems and directives of his employers, and how the military-industrial complex shaped the evolution of firms, sectors, and regions.[90]

Under William, from 1873 to 1886, Midvale Steel became a model second industrial revolution firm. Powerful men with large capital built the corporation and ultimately employed thousands of workers. With customers of national stature, they tied the growth and prosperity of the firm, and of the region, to national capitalism, especially the railroads and an expanding military. At the plant, they instituted scientific production through a regime of laboratory and shop floor research and development. Reforms in the organization of production gave rise to the new science of management. These patterns would characterize American industry and economic development through most of the next century.

CHAPTER 6

BUILDING THE SCIENTIFIC CITY

In an 1885 article in the *Journal of the Franklin Institute*, Coleman Sellers laid out his vision for the science of building cities. Machines produced in the workshop could improve urban life, he argued, only if the city was designed to accommodate them. "Great results followed in practice when a scientifically-constructed and well-laid road was ready for the locomotive," he observed. "Let us have smooth well laid streets, too," so that streetcars may move people efficiently between zones of home, work, and leisure. Transit, heat, light, and water were matters of civic importance, public goods that must be delivered efficiently. "The public are warranted," he opined, "in objecting to unsystematic and ill-advised engineering enterprises being conducted to their detriment and delay."[1]

For three decades following the Civil War, the Sellers firms shaped the built environment more intentionally and profoundly than in earlier generations. The tool works expanded production of railroad turntables and cranes, and William's rationalization of machine design influenced the ways engineers and architects approached design and construction in other endeavors and at larger scales. Edge Moor innovated in the manufacture of structural steel systems for bridges, tall buildings, and elevated subway lines that shaped cities and infrastructure systems from New York and Chicago to the West Indies and Australia. In the private consulting practice he established after leaving Sellers & Co., Coleman was best known as the lead engineer for the Niagara Falls power project.

This undertaking pioneered one of the key technologies that made possible the dispersed urban landscapes that Americans came to know in the twentieth century.

A scientifically designed metropolitan environment served the class and political interests of elite industrialists, and was therefore about power as much as technology. The Sellerses' capital and networks allowed them to undertake large civic projects and private real estate ventures in the first place. Deploying scientific expertise gave them another form of authority and sometimes a voice in public affairs. Their plans promised to advance their struggle against machine politicians to determine the course of urban and economic development, though the machine's municipal engineers also claimed science on their side. The environmental reforms and real estate investments the Sellerses pursued also aimed to influence the working classes, including their own employees.

In their urban and suburban development projects, William and his allies applied emerging theories of urban planning based in social and environmental science to restructuring regional land use and concomitant patterns of class. In the inner city, they established an early slum clearance organization that redeveloped one of Philadelphia's most impoverished alleys. For the elite, the Sellerses helped underwrite multiple projects of residential suburbanization. They also had a hand in the suburbanization of manufacturing and the working class in this era, mainly in the company town they built at Edge Moor. Through these projects, as well as in civic institutions and elite movements for parks, public art, and the removal of nuisance land uses, they strove to guide urban development and compete with the elite of other cities through their built environment. They also advanced critiques of industry in the urban environment and helped relocate it as they reorganized metropolitan land-use patterns.

Coleman and especially William engaged in virtually all sides of building a "scientific" second industrial revolution city. This work was the physical manifestation of their Progressive urbanism and illustrates further roles engineers played in multiple strands of early professional neighborhood, regional, and environmental planning.[2] It complemented and often tried to implement their ideas from the Social Science Association and other reform associations. For the Sellerses, rationalizing the built environment meant creating cities that reflected their social and political visions as well as their sensibilities as engineers and manufacturers.

Manufacturing Industrial Environments

In the post–Civil War era, William's businesses became increasingly focused on manufacturing the built environment of fast-growing industrial cities and regions. The metalworking and testing tools developed at Sellers & Co. made possible the standardization of steel parts produced by Edge Moor Iron for bridges and buildings, which in turn enabled the erection of stronger, longer, and taller structures. The massive turntables and cranes made by Sellers & Co. and Edge Moor aided railroad and streetcar companies, manufacturers, and shipping facilities in coordinating the movement of people and materials within and between cities. The boilers turned out at Edge Moor helped heat and power factories and office buildings. In functional as well as aesthetic terms, William and his companies influenced the design, development, and material form of the modern metropolis.

Mechanical engineers and historians have traced the equation of form and function in machine design principally to William. Engineering historian Joseph Roe wrote in 1915 that he "was among the first to realize that red paint, beads and mouldings, and architectural embellishments were false in machine design." Rejecting this tradition of prettying up machines, he "introduced the 'machine gray' paint which has become universal; made the form of the machine follow the function to be performed and freed it from all pockets and beading."[3] In his 1905 obituary of William, engineer Frederick Halsey explained these methods "were soon followed by other designers, and it is safe to say that, so far as modern machines are better in this respect than those of half a century ago, the result is very largely due to the influence of Mr. Sellers' work."[4]

In his address at the close of the Franklin Institute Exhibition of 1874, Coleman translated these mechanical principles into a broader theory of aesthetics. "It is an uneducated taste that finds satisfaction in brilliant colors only, or seeks to beautify uncouth forms by gorgeous paints," he averred. Instead, "a higher culture fashions forms to suit the purpose for which they are designed, and colors them in subordination to their uses and surroundings." For machines, this resulted in "a new order of shapes, founded on the uses to which they are to be applied and the nature of the material of which they are made . . . and the flaunting colors, the gaudy stripes, and glittering gilding has been replaced by this one tint, the color of the iron upon which it is painted." This new paradigm of machine design was more than a regional style or a quirk of engineers' profession. Rather, it revealed the rational ethos of industrial America. As Coleman put it, "That somber tint is no indication of any Quakerish objection to bright colors, but indicative of a higher culture and more refined taste."[5]

This aesthetic extended beyond the factory, influencing the work of architects and civil engineers. The mechanical drawing practices of firms like Sellers & Co., as well as some of their draftsmen and other students from the Franklin Institute's drafting classes, made their way into architecture firms.[6] The design of late nineteenth-century buildings, especially factories and railroad stations but also other edifices, adapted the mechanical aesthetic and the technological systems of Sellers and other engineers.[7]

This was evident, for example, in the two principal buildings of the Centennial Exhibition, the Main Building and Machinery Hall, which architect and civil engineer Joseph Wilson, William's Franklin Institute colleague, designed in the shape of immense foundries or train sheds. Amid the cash shortage following the financial panic of 1873, the exhibition's backers had little to spend on their construction. Sellers and Wilson responded with a combination of business and design strategies for these vast structures. Sellers & Co. supplied the shafting to transmit power from the great Corliss engine, the fair's most visited attraction, to the exhibits in Machinery Hall. Anticipating the modern pattern of the equipment lease, the firm submitted a reduced bid in return for ownership of—and the opportunity to resell—the shafting at the close of the fair.[8]

The buildings' iron structure was also reusable. With uniform parts from Edge Moor, and with the help of its engineers, Wilson designed the two halls in such a manner that they could be dismantled and their materials "readily disposed of in the market after the exhibition is over."[9] In 1881, after serving as an exhibit, meeting, and recreation facility for five years, the Main Building was purchased by an oil company executive. He employed its iron columns and trusses in the building boom going on around his headquarters in northwestern Pennsylvania, then the "oil capital of the world."[10] Edge Moor enabled the creation of flexible industrial environments with standard parts.

It was this opportunity to build a rationally manufactured landscape that led the Sellerses to establish the firm in 1868.[11] "The introduction of iron and steel into the construction of such engineering works as bridges, roofs, buildings, etc.," a later Sellers & Co. treatise declared, "has necessitated the establishment of manufactories devoted exclusively to this comparatively new industry, and in which the designs of engineers can be constructed in an accurate and inexpensive manner."[12] They founded Edge Moor mainly to make wrought iron and steel for railroad bridges, a logical outgrowth of Sellers & Co.'s product line for their core customers.

As the story of the Centennial Exhibition suggests, they soon branched out into the manufacture of other building systems. In 1878, William purchased the U.S. rights to make the Galloway boiler, a British invention that an Edge Moor circular called "the most economical and efficient steam generator now made."[13] The firm thus helped bring heat and power to American factories and office buildings.

The growth of these novel product lines necessitated a "comprehensive hydraulic plant," including many new machines patented by Sellers & Co., which sold to other structural steel makers as well.[14] One of William's goals for Coleman's European trip in 1884 was to find stronger pumps to run Edge Moor's hydraulic equipment. He wrote his cousin, "There is no question but that our product per man largely exceeds any other works in this country, but our expenses neutralize this and must continue to do so until we can concentrate more business here." Workmanship and materials were not the issue. He maintained, "No one can make eye bars to compete with ours and now that we can sell steel bars for less than iron, our advantage is increased." But more efficient use of power "in our riveting would largely diminish expenses in that department, so keep your eyes open to the best pumping engine to work at 600 pounds water pressure," he counseled; "if there is anything better abroad than we have, I shall be glad to know about it."[15]

Hydraulic tools permitted Edge Moor's bridge department to make stronger bonds between the parts of a bridge. This avoided the weak joints created by steam or hand riveting, which could cause disaster when they failed.[16] By the 1880s Edge Moor developed an early form of assembly-line production using machine tools specially designed for each function.[17] Testing machines made by Sellers & Co. to analyze the strength of steel bars and beams under pressure helped ensure the safety and durability of bridges prior to their erection. With these tools, claimed Joseph Roe, at Edge Moor "the building of bridges [was] . . . put upon a manufacturing basis" distinct from the craft traditions of the construction industry.[18]

Edge Moor Iron, Carnegie's Keystone Bridge Co., the Union Bridge Co., and the Philadelphia area Phoenix Bridge Co. together made American bridge *manufacturing* distinct from European bridge *building*. Coleman made this clear in a somewhat veiled juxtaposition of Scotland's foremost steel bridge, which he witnessed under construction, against Edge Moor's bridge over the Susquehanna River for the Pennsylvania Railroad, which at the time was the world's longest double-track railway bridge. At the Queen's Ferry bridge near Edinburgh, he was struck by "the acres of shops

Figure 18. Edge Moor's Susquehanna River Bridge for the Pennsylvania Railroad, c. 1885. Wilson Brothers & Co., *A Catalogue of Works Executed* (Philadelphia: J. B. Lippincott, 1885). From the collection of George E. Thomas.

that were erected at the place to construct this enormous work from the plate-iron and bar-iron as it came from the rolling-mill." He disparaged the "shops and . . . machinery that would become useless and must be removed upon the completion of the structure."[19]

By contrast, Coleman observed, "Iron and steel bridges of good design and of undoubted strength are being erected [in America] that are made in large manufacturing establishments far removed from the site of erection and are arranged so as to require no [machine] work in assembling the parts and putting up the structure." The Susquehanna bridge "was erected in six months from the date of order without any hand work, such as chipping or altering, and the progress of the erection did not interfere with the travel over the road." Manufacturers like Edge Moor, he maintained, made it possible "to do the work more rapidly and more thoroughly than is possible in shops built for the occasion and to which in out-of-the-way places skilled workmen must be carried and domiciled during the term of erection."[20]

Edge Moor's own location also illustrated the ways in which industrialists reshaped the geography of manufacturing in the late nineteenth century. Unlike Sellers & Co., which enjoyed close proximity to customers and a large pool of skilled labor in its central city location, the structural steel company sought the benefits of industrial decentralization. The Sellerses located Edge Moor's plant outside of Wilmington. With freight connections to the Reading, Baltimore & Ohio Railroad and the Pennsylvania Railroad and a mile of frontage on the Delaware River, it had easy access to national and international markets. This site, which

extended from the mouth of the Christiana River to the DuPont pow-
der warehouses, was large enough to accommodate Edge Moor's growth.
Its relative isolation also allowed for greater control over its workforce.

Edge Moor was an important part of Wilmington's post–Civil War
industrial expansion, and it benefited from the networks of firms and institutions in this city as well as Philadelphia. The Sellerses founded the company at a time when the city's factories were booming, the local Chamber
of Commerce was energized, and state and municipal authorities were
investing in its port and transportation infrastructure.[21] As a relatively small
place, though, Delaware offered industrialists greater influence over economic policy and public investment than they held in Pennsylvania, where
more diverse and powerful interests competed across larger territories. In
order to procure skilled labor that could remain in some ways separate
from the city's existing working class, William went to the iron region of
Lancashire, England, to recruit craftsmen. He housed many of them in the
company-owned village he developed adjacent to the factory.[22]

The Sellerses' family ties in and around Wilmington, together with
their capital, made them immediate players there. In addition to William
and John Jr.'s sister Mary's marriage into the Bancroft family, their sister
Frances wed local iron merchant Eli Garrett. He assembled most of the
land for Edge Moor, joined his brothers-in-law as a partner, and served as
the company's secretary and treasurer.[23] The family of William's first wife,
Mary Ferris, owned the hill overlooking Edge Moor, where the couple established their country house in 1868. The ties of the Bancrofts, Garretts,
and Ferrises to other Quakers, merchants, and industrialists in the area
helped the Sellerses build social and business networks that would serve
their company's needs for labor, infrastructure, and professional services.
Another family member, William and John's younger brother George,
also joined the firm, moving to Delaware to serve as its manager.[24]

Much of the money to build Edge Moor came from Sellers & Co.,
which by the turn of the twentieth century loaned nearly a million dollars to its sister company. With an 1875 mortgage from the Pennsylvania
Company for Insurance, the firm leveraged additional capital to expand its
physical plant. At the time, seven years after its founding, according to the
mortgage contract, it consisted of "Puddle and Rolling Mills, Producer
House, Engine Houses, Boiler Houses, Machine and Forging Shops, Store
room, Laboratory, dwelling houses, offices, sheds, school houses, tools,
implements, machinery and fixtures."[25]

Edge Moor grew quickly, as it took on some of the biggest bridge projects expanding and linking cities in late nineteenth-century America and

abroad. By the time of Eli Garrett's death in 1886, between five and eight hundred men worked at the plant, depending on the volume of orders. The company employed some two hundred more erecting bridges, buildings, and boilers for customers throughout the United States and abroad. They manufactured multiple spans across the Mississippi River. Their Kentucky River Bridge for the Cincinnati Southern Railroad was the highest railway bridge in the nation. Their Schuylkill River Bridge in Philadelphia brought the Pennsylvania Railroad into the heart of the downtown. Edge Moor built elevated railways for mass transit companies in New York City and Kansas City, Missouri. The firm partnered with Joseph Wilson on numerous railroad bridges in the United States and Australia, including the first iron rail viaduct in Oceania. They also collaborated on one of the largest market halls in the West Indies, in Demarara, Guiana.[26] Through their international customer bases, they disseminated iron and steel construction methods to other parts of the industrializing world.

Recognizing the specialized nature of this line of business, in 1888 the Sellerses incorporated the Edge Moor Bridge Works. Edge Moor Iron continued to make boilers while, as the Dun credit agency reported, the bridge works produced "iron and steel bridges, viaducts, elevated railways, marine piers, iron and steel work for buildings and roofs, and . . . wrought iron turn-tables."[27] The bridge company hired a string of graduates from Rensselaer Polytechnic Institute, the leading civil engineering school in the nation. Their output ranged from small bridges over creeks in Delaware to longer spans across the Monongahela River at Pittsburgh and the Columbia River in the Pacific Northwest.[28] For the Manufacturers' Building at the World's Columbian Exposition in Chicago in 1893, Edge Moor Bridge erected the largest steel roof arches yet built. Like the 1876 Centennial exhibit halls and the market at Demerara, this building demonstrated American engineers' newfound capacity to build tall structures containing huge expanses of interior space for commercial, manufacturing, and recreational activity.

Edge Moor Iron's boilers drove the steam heating and power systems of factories, power plants, and office buildings. The company supplied boilers to major manufacturers such as Proctor & Gamble outside Cincinnati; the Pacific Mills in Lawrence, Massachusetts; the U.S. Mint and Midvale Steel in Philadelphia; and Alexander Holley's steelworks at Troy, New York.[29] Customers among utilities included Metropolitan Telephone and Telegraph of New York, Chicago Edison, and the waterworks of Wilmington, Philadelphia, and Cincinnati. Institutions ranging from the State Hospital in Rochester, New York, to the University of Pennsylvania and

Figure 19. Edge Moor's market house at Demerara, Guiana, under construction, 1882. Wilson Brothers & Co., *A Catalogue of Works Executed* (Philadelphia: J. B. Lippincott, 1885). From the collection of George E. Thomas.

Philadelphia County Prison ran their physical plant on Edge Moor boilers. The firm's boilers also distributed heat and light in many of Philadelphia's most prominent late nineteenth-century office towers and department stores.[30]

Like Sellers & Co., the firm gained a strong reputation for its product. In an 1893 testimonial, the president of the Galveston Rope & Twine Co. in Texas quipped, "I am inquiring for some anti-fat medicine to keep" the boiler room fireman "of a proper size to get through the manhole" of the unit he had purchased from Edge Moor. Like the "Maytag man" a century later, this customer related, "he has so little to do in running the boiler." The somewhat higher cost of Edge Moor's product, he claimed, "is more than made up in the savings in fuel and repairs."[31]

However, unlike the Bush Hill machine tool works, Edge Moor struggled to make a profit. In 1889, Dun reported that the company "has not been much of a financial success," though the rating agency considered its credit sound since the Sellerses owned nearly all its stock. They continually poured more of their money into upgrades to its plant and equipment.[32] Edge Moor could boast a long list of impressive engineering feats, but as a pioneer in this sector the firm sustained high costs in extensive research and refinement of production. Rather than an enduring competitive advantage, however, rationalization helped turn bridges into another standardized commodity as bridge building became highly competitive.[33]

Unfortunately, Edge Moor's most disastrous contract was also its most public one, the steel superstructure of the Brooklyn Bridge, completed in 1883. The project's notorious delays, by no means due to Edge Moor alone, coincided with the continued Broadway success of the play based on *The Gilded Age*. This opened the door for the New York press to make William the brunt of jokes about the bumbling "Colonel Sellers" and his attempts to build a bridge.[34]

Beyond the drama, the Brooklyn Bridge—the first steel span between the largest and the fifth-largest cities in the nation—illustrates some of the challenges Edge Moor encountered in practice. The bridge's specifications changed substantially during construction, magnifying the already complicated management of such a large-scale undertaking. More important, soon after Edge Moor won the job, an upswing in the price of steel as well as labor rendered it an unprofitable contract. These sorts of problems affected essentially all late nineteenth-century makers of heavy machinery, who with other manufacturers struggled in the context of general deflation in this period.[35]

William's institutional strategies would not make Edge Moor very profitable, though they did advance engineers' involvement in urban planning and infrastructure design. Edgar Marburg's engineering research and Edmund James's study of public utility administration at Penn aimed to establish the standards by which municipal utilities and departments of public works operated. Political reform associations the Sellerses joined pushed a similar point in their call for professionalization of civil service.

The Franklin Institute also tackled technical questions of import for urbanization. Its committee on roads supported research to improve paving technologies. The institute's lecture series in 1886, to take just one year as an example, included talks on "Elementary Problems of Bridge Construction," "The Development of Transportation Facilities in the United States," "Natural Gas," "Water-Supplies of Cities," "Sanitary Care of Cities," "Sanitary Plumbing," and "The Purification of Water-Supplies."[36]

As the latter subjects implied, the Sellerses and their colleagues concerned themselves with much more than the mechanical and chemical engineering systems to build cities of steel. From the small scale of Sellers & Co.'s machine tools to the emerging field of sanitary engineering, they sought to make the design of the built environment a more rational and scientific process—one over which they would have greater influence. At Edge Moor, they devised new technologies and modes of building big structures, even if they struggled to make it a viable business. They also necessarily grappled with some of what economists call "negative externalities" of urbanization and industrialization, namely pollution and related health problems. They manifested these commitments again through a set of public and private institutions that fought for the power to define how the city grew.

Engineering Environmental Planning

Historians trace the immediate origins of the city planning profession to the parks and public health movements of the late nineteenth century.[37] Much of this story centers on a few great designers like Frederick Law Olmsted. Engineers like the Sellerses also played active roles in what would later be termed environmental planning, including design, development, and environmental management functions. They sought to coordinate decisions about the natural environment, public health, working landscapes, and urban growth.[38] Defining the terms of debate and investment in environmental reforms enabled the Sellerses and their allies to push their class and business interests, sometimes to discipline the Republican machine, and more broadly to confront some of the region's foremost social and health concerns.

This was mainly a civic enterprise. Beyond the elite associations the Sellerses and their associates joined to promote parks and public art, William wielded state-sanctioned authority as a member of the Fairmount Park Commission, remaking a large part of Philadelphia and redefining portions of its manufacturing landscape. John Jr. led one of the major citizens' protests of this era against industries that threatened the water and air quality of the central city. In the process, the Sellerses helped shape city dwellers' conceptions of the proper place of industry in the urban environment.

The most important focus of these debates was water, arguably the resource most threatened by the growth of industrial cities. Fairmount Park would be the chief instrument for protecting Philadelphia's water supply. William and fellow commissioners did this by "scientifically" managing the Schuylkill River watershed.

Fairmount Park extended the powers of the state, and initially of the Sellerses and their allies, over the metropolitan landscape. As early as 1854, just after the Consolidation Act passed, John Jr. joined a long list of petitioners to city council urging the creation of a municipal park along the Schuylkill.[39] In 1867 the state legislature created the Fairmount Park Commission, a public body appointed by the city courts and possessing broad authority over the region's rivers, creeks, and adjacent lands. By the early twentieth century, it would acquire roughly one-tenth of the city's land area. In the process, it removed industries from substantial parts of the Schuylkill River and several creek valleys.

The first commissioners came from the ranks of civic reformers and engineers, including Mayor Morton McMichael, real estate magnate Eli Kirk Price, merchant John Welsh, and a variety of engineers, among them William Sellers. He served on committees for Land Purchases and Damages, Plans and Improvements, Finance, and Superintendence and Police.[40] Through these and other ad hoc committees, he helped plan and implement major environmental engineering reforms.

In July 1867 the commission referred the analysis of the water supply and its purity to a committee composed of William and other engineers to establish a scientific basis for decisions about land acquisition for the park. In October, they returned a report assessing the chemical properties of the Schuylkill water. They compared Philadelphia's water system to those of Detroit, New York, Quebec, Geneva, Glasgow, Manchester, London, Paris, and other cities. The Schuylkill measured favorably against the rivers of these rival industrial cities. Still, the committee argued for vigilant regulation of manufacturers upstream. The dumping of dyes and other waste into the river, they proclaimed, was "a wanton outrage upon common decency." In the view of this group of engineers, pollution could "be restrained without affecting in the least the business of the parties employed in manufacturing on the banks of the stream."[41]

The solution, the committee concluded, lay in two main strategies. For the denser concentration of mills in Manayunk, Falls of Schuylkill, and Brewerytown, the city should build sewers to carry the waste of upstream textile mills, a major chemical plant, and breweries to a point below the intake of the waterworks. They recommended confiscating manufactories and other properties along long four miles of the Schuylkill River where industry was sparser and seven miles of old, smaller mills along its main tributary in the city, the Wissahickon Creek.

Committee members exhorted their colleagues to use the powers the state had granted them to make these large acquisitions of property. Despite consolidation more than a decade earlier, the committee submitted, "The legislative powers vested in the City have never yet been exercised to any considerable extent."[42] With the commission's mandate to protect the water supply, they must now revise land-use patterns by actively acquiring private property, whether or not the owners wanted to sell.

The following year, William and fellow members of the Committee on Plans and Improvements took up the crucial question of "the boundaries within which it is desirable and proper that the . . . Park should be contained." The answer to this question was partly a scientific assessment of how to best prevent pollution from washing into waterways. It was also a function of real estate development. The committee acknowledged that "the owners of the land are watching with interest the steady approach of improvements that are pushing toward and will presently reach and surround them."[43] Some of those owners, most notably Eli K. Price, sat on the commission.

With boundaries well beyond the built-up areas of the city, the park would structure much of Philadelphia's future growth and help define its status among the world's cities. The commissioners estimated that the city's population "in twenty years, will probably be a million and a half." Thus, "with passing time our overflowing population shall have embosomed these spacious grounds with the homes of the people of a vast and prosperous City" built around the park.[44] Viewing Fairmount Park as an instrument of Philadelphia's competition with other cities, the commissioners resolved that Morton McMichael's trip to Europe in 1869 should include visits to the continent's great urban parks, "soliciting and obtaining such information with reference to their laying out and adornment and the government and administration."[45]

William and his colleagues set about engineering the roads and recreational landscape of the park. Through the Centennial Committee he headed, the commission leveraged public and private investment in the centennial to further study European park planning, pave roads, and construct restaurants and zoological gardens.[46] To pave the park drives, his fellow commissioners called on William "to ascertain and report upon the best kind of Steam Roller."[47] Together with fellow "gentlemen educated as Engineers," William studied "in what manner the safety and convenience of the public in crossing the River should be provided for while" new bridges over the Schuylkill were under construction.[48]

Most improvements aimed at what the commissioners termed "popularizing the park," creating the spatial framework for different constituencies to inhabit it. In their 1871 annual report, writing in the third person, they acknowledged that "while providing as rapidly as circumstances would permit for carriage roads and bridle paths, they have been equally sedulous to secure facilities of access and convenience of movement for those who do not indulge in expensive pleasures."[49] This included bringing streetcar lines into the park and, at William's suggestion, "the improvement of the Springs" for drinking fountains, "proper Urninaries," as well as horse troughs for the carriage and equestrian set.[50] The commissioners viewed civil behavior among the working classes as critical for the park's success in attracting an elite constituency and boosting Philadelphia's reputation as well as adjacent land values.

Through the private Fairmount Park Art Association, established in 1871, the commissioners enlisted other elite Philadelphians in creating a civic landscape in their own image and in advancing social reform through the park's environment.[51] William, John Jr., and Coleman all joined. The association aimed "to erect statues, busts, and other works of art, to the honor and memory of eminent persons and their deeds, in order that their good example and influence may be more permanently exerted in fostering a spirit of emulation and ambition in our people."[52] Such projects and ambitions were characteristic of the parks and other urban reform movements led by elites across metropolitan America.

William, Morton McMichael, and some of their colleagues also envisioned the Fairmount Park Commission as a vehicle for political reform. Its charter from the state and the appointment of commissioners by judges, as opposed to the mayor or council, gave it some autonomy from machine politics, at least in its early years. By 1877, when the city judges denied William reappointment, the machine had effectively infiltrated the institution. During his tenure, however, he achieved some victories for fiscal prudence and against corruption. At the commission's inception, he introduced resolutions to govern accounting and contracting. For a time he helped prevent the park police force from becoming a bastion of patronage.[53]

As its initial report on the water supply averred, the commission's greatest influence over Philadelphia's development derived from its powers to buy private property for public use, with or without the owners' consent. In the spring of 1868, the Committee on Land Purchases and Damages, which included William, met "twice a week," committee chair Eli K. Price informed the press, "every Tuesday and Friday mornings, to hear claimants, besides giving other time to view the properties and make their estimates."[54] They assembled vast tracts of land, buying up the valleys of the Schuylkill River and Wissahickon Creek.

This land acquisition and the sewers built to divert waste from remaining manufacturers achieved their predicted results, for real estate as well as the water supply. By 1877, Price could report an addition of "500 per cent. to our capacity to store water, which, so far as an adequate supply of water will do it, will insure the City's future safety from any great

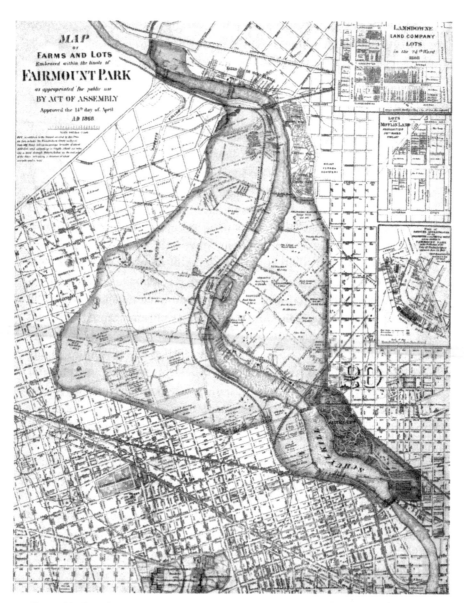

Figure 20. Plan of Fairmount Park showing farms and lots acquired by William Sellers and his colleagues in the Committee on Land Purchases and Damages, 1870. *Commissioners of Fairmount Park: Map of Farms and Lots Embraced within the Limits of Fairmount Park as Appropriated for Public Use by Act of Assembly Approved the 14th Day of April, A.D., 1868.* Map Collection, Free Library of Philadelphia.

conflagration." In addition, he noted a marked rise in property values and home building on the borders of the park, especially near the Centennial Exhibition fairgrounds and the railroad suburbs-in-the-city bordering the Wissahickon Valley. Price estimated "we shall soon have value added to our real estate and taxable resources more than commensurate with the purchase money of all the Park, and perceive the effect to increase in the future indefinitely."[55]

Land acquisition for the park reorganized the geography of manufacturing along an eleven-mile swath of the city. The Committees on Plans and Improvements and on Land Purchases and Damages negotiated and enforced the closure of the paper mills and tanneries of the Wissahickon across several years. They took breweries, icehouses, metal manufactories, and oil refineries along the Schuylkill above the waterworks, including one owned by machine politician and Standard Oil partner William Elkins. This effectively confined oil refining and other big industries to the lower Schuylkill below the waterworks. Though some breweries were demolished, the firms of Brewerytown that remained, up the hill from the river, benefited from a cleaner and more secure source of water. The commissioners also blocked the construction of new railroad lines through the park, hid existing tracks with fences and "impervious hedges," and buried sewer lines or concealed them behind picturesque stone archways and walls.[56]

Around the waterworks at Fairmount, at the main entrance to the park from the downtown, they completely made over land-use patterns. Already by 1872 the park's leaders could announce, "railway tracks have been taken up; furnaces, foundries and iron mills have been removed; huge ditches and broad canals have been obliterated." They demolished entire blocks, as "cluttered streets and dusty highways have been vacated; and in place of these are seen rural sights, already beautiful, and soon to be rendered tenfold more so."[57] Although they cleared areas least built-up with industry and took relatively few homes beyond a large group of old estates, they established a precedent for replacing industrial landscapes with public spaces in the name of health and civic improvement.

Some of the rhetoric and actions of the park commissioners signaled that the place of manufacturing within the city was coming under increasing scrutiny and regulation. This manifested itself in public institutions like the park commission and in private forums such as the art association, which generated similar language about the civilizing effects of an escape from the city's most industrial landscapes. As long as industrialists like the Sellerses continued to lead institutions like the park commission, they were generally protected from the list of industries targeted, though the family mill on the Wissahickon was shut down.

As homeowners, the Sellerses and other elite Philadelphians advanced critiques of industry that addressed perhaps the greatest flashpoint for urban environmental reform in the nineteenth century, namely meat. Animal agriculture played vital roles in cities' food supply, manufacturing and retail sectors, and waste management, but it also presented some of the greatest threats to public health and the comfort of neighborhood environments (and consequently real estate values). This inspired the Sellerses to take on their machine tool works' most important customer. Many wealthy men opposed the Pennsylvania Railroad's 1874 proposal to erect an abattoir on the west bank of the Schuylkill just below the waterworks, a twenty-one-acre stockyard capable of processing some twenty-five thousand pigs and cattle per day. But it was John Jr. who would be the lead plaintiff in the legal fight against it.

When the newspapers first reported the abattoir plan, "a large class of the community, citizens and tax-payers, owning or representing many millions of dollars in real estate in the immediate neighborhood," issued a "remonstrance" to the company. "To locate a nuisance of this description in a spot which is now close to the thickly built portions of the city," they complained, "would be to imperil the health and comfort of the citizens, and to greatly depreciate the value of property along the whole western portion of the old city." Prevailing winds would "carry the offensive effluvia to the handsomest improvements and most highly taxed dwellings in the city, while the drainage, swept backwards and forwards with the tide, will convert the Schuylkill . . . into a source of dangerous infection."[58]

The location of this facility would be unlike other stockyards and slaughterhouses, which concentrated in working-class districts of nearly every American city. This abattoir would be a few blocks east of the University of Pennsylvania's new campus and the affluent new residential neighborhoods of Powelton, home to Coleman and John Jr., and Spruce Hill, where the Drexels and Clarks resided. Just across the river in the neighborhoods around Logan Square, where William lived, and especially surrounding Rittenhouse Square were town houses of the city's elite—they were in the path of the prevailing winds. In addition to residents, the doctors and administrators of hospitals on both sides of the river supported the subsequent suit against the railroad.

Giving testimony in the case of *Sellers et al. v. the Pennsylvania Railroad Company*, witnesses for the complainants as well as the defense deployed

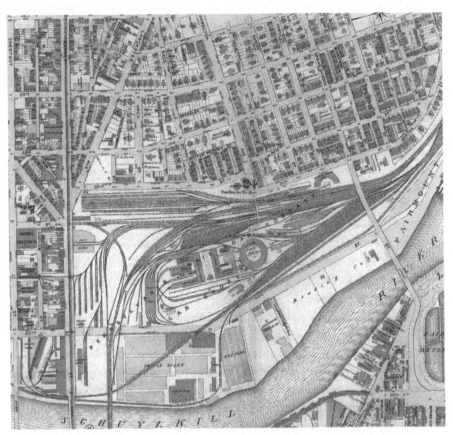

Figure 21. Atlas of the area around the Pennsylvania Railroad's abattoir. At the top is the Powelton neighborhood where Coleman, John Jr., and other members of the Sellers family resided. Geo. W. and Walter S. Bromley, Civil Engineers, *Atlas of the City of Philadelphia* (Philadelphia: G. W. Bromley & Co., 1901), plate 9. Map Collection, Free Library of Philadelphia.

scientific critiques of industry in the urban environment. The plaintiffs called upon experts including doctors, public health officials, and politicians from other cities. They charged that the abattoir would "add to the already very great mortality in this city among infants, . . . retard recovery in cases of disease in its neighborhood," and "increase the present dangerous and malarious condition of that river in summer months."[59] They lined up real estate brokers to argue that it would bring down property values in adjacent neighborhoods. These "scientific" claims about health and property would be repeated in Progressive campaigns to reform Western nations' cities across the ensuing decades.

The railroad countered with witnesses who claimed the proposed butchering was no worse a threat to public health and welfare than other manufacturers. "Pig Iron" Kelley, a great supporter of the railroad in Congress, had lost his sense of smell three years earlier. However, he testified that the proposed facility, which was to receive all of its animals by rail, "would be an absolute relief to us from both the driving of cattle [in the streets], and the odors from ill-managed and ill-drained private slaughterhouses."[60] City Surveyor Hudson Shedaker charged that immediately across the river the city gasworks emitted sulfur that was "infinitely more noxious and prejudicial to health."[61] Several blocks north were "extensive lime kilns, where oyster shells are burned, which occasion offensive odors nearly as great," and just beyond was the Spring Garden Gas Works.[62] In the end, there was little legal or political support for the plaintiffs; the railroad won the case, and the abattoir was built at the proposed site along the river. Although its elite neighbors lost their case, it did help remove animals from the streets, a dramatic change in the public environment of late nineteenth-century cities.[63]

Engineers at the Franklin Institute addressed the public health issues raised by Fairmount Park and the abattoir case, too. In the 1880s they established a separate department for sanitary engineering, "with special reference to Household Plumbing and Drainage and the connecting private and public sewerage." Another new department tackled ventilation, "with the view of correcting wastage of fuel for heating purposes; the indrawing of damaging currents of cold air; and the injurious effects produced in some instances in the sick chamber and Hospital wards, by the upward draught of noxious matter."[64] These would be prominent topics in the first national conferences that defined city planning as a profession, and engineers understandably led this most scientific side of the field.

In the decades after the Civil War, the Sellerses and their colleagues and competitors elaborated scientific interventions and discourse of urban environmental reform. The movements and institutions in which they participated articulated plans and sometimes gained the authority, especially in the case of Fairmount Park, to implement major changes in the geography of public space, infrastructure, residence, and manufacturing. The abattoir case reveals how elite interests allied to protect real estate investments and the livability of urban neighborhoods. Together, these projects illustrate how capitalists elaborated critiques of industry in the urban environment. More broadly, the development strategies and scientific critiques employed by the Sellerses and their allies articulated a Progressive agenda for environmental planning and public health. Through this vision, they also rationalized and reified the restructuring of land use and class across the metropolis.

Making the New Urban Order

The Sellerses' technologies, institutions, and real estate investments converged to shape new regional growth patterns. They helped build an increasingly vertical central city, around which they and other capitalists created new configurations of home and work. In the late nineteenth century, William and John Jr. chose to put large sums of their capital into real estate and construction. The places they planned and built represent a cross-section of the new metropolitan geography.

At the region's center, the technologies pioneered by Edge Moor and its competitors enabled the rise of a new downtown, including the earliest skyscrapers. Just four blocks south of Bush Hill, in 1871 construction began on Philadelphia's new City Hall. The Edge Moor Bridge Works supplied iron beams for the interior metal frame of City Hall's office blocks and tower, designed to be the tallest building in the world (though it took three decades to complete and was surpassed by other projects).[65] At least in the early 1870s, the firm was not above doing business with the Republican machine that drove this extravagant project.

Across Fifteenth Street from City Hall, at the head house of the Pennsylvania Railroad's Broad Street Station, the Sellerses' civil engineering colleague Joseph Wilson designed the world's first skyscraper in 1881. Architectural historians have erroneously credited Chicagoan William Jenney with this innovation.[66] It was bridge companies that made skyscrapers possible, technologically, together with civil engineer colleagues like Wilson and Jenney who helped adapt their structural iron and steel systems to tall buildings. Historian of technology Thomas Misa points out that all the major "Chicago School" architects employed bridge engineers, including Edgar Marburg, who worked first at Keystone Bridge, then Phoenix Bridge, and later for Edge Moor.[67]

If skyscraper skeletons were like Edge Moor's bridges turned on their ends, Coleman described their elevators in a similar vein. They rendered "the upper rooms of high buildings more desirable for office purposes and residences than even those nearer the ground. Such cities as New York," he observed, "have through their hoisting systems obtained vertical railroads and rapidly-moving cars that have thrown the staircases in the same buildings into disuse."[68] These technologies structured new forms of white-collar work, retail, residence, and also manufacturing.

Many manufacturers established corporate offices downtown, away from their plants but close to their lawyers, financiers, and sales agents. By 1890 the Sellerses' Millbourne Mills Co. set up an office on Broad Street just south of City Hall.[69] They took advantage of proximity to the offices of wholesale merchants, banks and trust companies supplying credit, and railroads with which they negotiated discounted rates for freight. With the aid of the telephone, the company's managers coordinated orders for western wheat and sales of flour to a national market.

The standard story about manufacturing in the late nineteenth century is that it exited the city. But it did so only in part, and before many heavy industries moved they played a significant part in the changes reordering central city environments. With a high premium on real estate and inner city districts increasingly built out, many factory owners were forced to abandon the old practice of expanding their plants onto adjacent properties. Instead, they pursued spatially intensive strategies to boost production.[70]

Into the twentieth century, Bush Hill remained home to three of the region's largest industrial employers: the Baldwin Locomotive Works, the top employer; Sellers & Co.; and Bement. In 1896 the U.S. Mint built its new plant adjacent to Baldwin and Sellers, confirming the district's continued importance as a metalworking center.[71] At the Baldwin works, Sellers & Co. installed two one-hundred-ton, high-speed, electrically driven cranes in 1890 that each allowed a single operator to hoist and carry the largest class of locomotives over other engines being assembled in the erecting shop. This $60,000 investment paid off in just two years, as the cranes quickened production and permitted the firm to cut sixty lower-skilled workers who previously lugged parts around this area of the plant.[72]

Bush Hill manufacturers also expanded vertically. In 1906, when a reporter from *Locomotive Fireman's Magazine* visited Sellers & Co., the injector department had "recently been moved into a commodious five-story building, where the whole business is carried on under one roof." It integrated production, testing, training, and marketing in a single structure. "A noticeable departure" from standards of factory design, the visitor reported, "is in the location of the foundry, core shops, metal stores, etc., which are placed on the upper floors."[73]

This altered the flow of materials, something Baldwin did in remodeling eight of its buildings around the same time.[74] At Sellers & Co., according to the reporter, "The heavy pieces are delivered to the second floor by elevator, while the small pieces are dropped to the third floor through a wrought iron chute, which lands the castings right where they are to be machined."[75] This pattern of assembling machinery as it made its way down floor by floor would enable heavy manufacturers to remain in the central city for decades, even as many large producers opted for single-story plants on sizable acreage in the suburbs.[76]

The residential trajectories of Bush Hill proprietors and workers show how home, work, and different social classes became increasingly separate as commercial and industrial land uses intensified.[77] In 1860, nearly everyone who worked at the tool works lived within a fifteen-minute walk. Coleman, John Jr., the Bements, and other Bush Hill proprietors lived in the town house neighborhood of Spring Garden just north of the factory district. Some of Sellers & Co.'s machinists continued to dwell in the back streets of factory districts, but they increasingly moved to the brick row house neighborhoods north of more affluent Spring Garden.[78]

Across the late nineteenth century, the development of railroad and streetcar suburbs enabled factory owners and the growing managerial class to live farther away from areas of large-scale commerce, manufacturing, and working-class housing. In 1866, William and John Jr. spent $23,000 to purchase an entire city block in the leafy streetcar suburb of Powelton, just across the Schuylkill River.[79] There, J. Sellers Bancroft, Coleman, John Jr., and their sons would build large stone homes surrounded by gardens. While most of the Sellers family resided in Powelton by 1870, the firm's lower-level managers and draftsmen would not leave for the suburbs until the 1880s. By 1890, however, several of them lived in the railroad suburbs-in-the-city of Germantown and Mount Airy.[80]

Although William and his first wife, Mary, acquired the Clifton estate in Delaware, they maintained their primary residence in central Philadelphia, moving in the late 1860s to a large brownstone town house fronting on Logan Square. Both his old Philadelphia home and his new one were just three blocks from his factory. His neighbors still included the Bement family and Baldwin superintendent Samuel Vauclain. The Bements bought several homes along the border between Bush Hill and Spring Garden, which it seems they may have rented to some of their mid-level employees.[81] However, the proprietors and top management of most Bush Hill firms would move away from the district by the end of the century. William would keep his home on Logan Square but spend more and more nights at Clifton.

Railroad directors and their bankers led the way in suburban residential development, partly since they could influence the location of new lines and stations, but manufacturers like the Sellerses also invested in suburbanization, sometimes for a broader spectrum of the region's population.[82] William and John Jr. used funds from their tool works to build elite, middle, and working-class suburbs. In addition to their enclave in Powelton and company town at Edge Moor, they bought up considerable tracts on the Philadelphia side of Cobbs Creek in the postwar decades, where streetcar lines and the decline of mills downstream from Millbourne promised to soon bring residential development for the middle class (the valley would become part of Fairmount Park at the beginning of the twentieth century). In Delaware County, they took advantage of the demise of water-powered mills along the Ridley Creek, joining with fellow directors of the Philadelphia, Baltimore & Washington Railroad and other capitalists to acquire five hundred acres in the early 1870s for an elite enclave where the creek met their line.

They envisioned Ridley Park as a model railroad suburb. Through the Fairmount Park Commission, William met landscape architect Robert Morris Copeland, whom he and fellow investors engaged to plan the "Park City of Ridley."[83] William's Saturday Club colleague J. B. Lippincott promoted the venture in his magazine as the only community yet designed that compared with the Chicago suburb of Riverside, which was laid out by Copeland's competitors, the preeminent landscape architecture firm of Olmsted & Vaux. The earlier garden suburb of Llewellyn in northern New Jersey was "on somewhat unmanageable ground, and designed without the adornment of lakes," the magazine noted. This made it inferior to Ridley, which "with its two or three large ponds, its cascades and streams, and its splendid view of the sail-whitened Delaware, will be *par excellence* the aquarelle or water-color picture in these different specimens of natural fine art."[84] Philadelphia's elite sought to compete with the bourgeoisie of other metropolises through landscaped suburbs as well as downtown skyscrapers and world's fairs.

Lippincott's advertisement presented a romantic, civic vision, targeting buyers accustomed to urbane amenities. It cast the landscape planner as an artist, "painting with real forests and real waterfalls," some of which likely spilled out of former millponds, though the magazine failed to mention them and instead implied the "park city" was carved out of its natural setting. "He takes from his palette a lake and sets it on the foreground, where yonder dank meadow and stagnant stream are lying; the lovely villas dip their reflections into it in a crescent. Behind their turrets and gables rise the dark velvet groves." Ridley Park's civic landscape included "the city square, the town-hall of dazzling marble; all the winding lanes and descending

RIDLEY—A LANDSCAPE-GARDENER'S CONCEPTION OF A PARK CITY.

Figure 22. Drawing of Ridley Park from *Lippincott's Magazine* in 1872, "a landscape-gardeners' conception of a park city," showing the sylvan landscape envisioned by landscape architect Robert Morris Copeland for his clients. *Ridley—A Landscape-Gardeners' Conception of a Park City*, from "Through William Penn's 'Low Countries,'" *Lippincott's Magazine of Popular Literature and Science*, September 1872, 257. Print & Picture Collection, Free Library of Philadelphia.

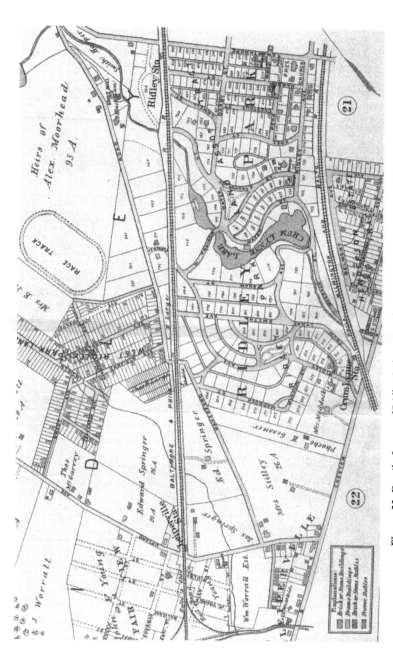

Figure 23. Detail of a map of Ridley Park as it was laid out and partly built by 1889, exhibiting the curved streets and large lots characteristic of early garden suburbs. From: J. L. Smith, *Atlas of Properties along the Philadelphia, Wilmington and Baltimore R.R., Baltimore & Ohio R.R., Philadelphia & West Chester R.R. and Part of Philadelphia & Reading R.R.* (Philadelphia, 1889), plate 23. Map Collection, Free Library of Philadelphia.

terraces somehow arrive at this public and splendid nucleus," which also included preselected sites for churches, schools, shops, and a hotel.[85]

Ridley's designer and developers adopted a broad complement of spatial and institutional strategies to guide its future growth, reflecting Progressive urbanists' holistic approach to suburban development that would soon proliferate across U.S. metropolitan regions. They sheltered it from the factory districts and working-class towns that dotted Delaware County. They regulated land-use patterns through open-space planning and the rules of its development association, which incorporated the town as an independent municipality with its own government. They created the Ridley Park Cold Springs Water Co. in 1889, which William headed, to pump water from the lake to a tower that supplied the homes. The *Boston Advertiser*, Copeland's hometown paper, celebrated the large areas around the brooks and ponds that were "given to the public," and by the early 1890s this became a popular place among Delaware County residents for summer boating and winter ice skating.[86] The designer reserved a twenty-acre park and a fifteen-acre cemetery for the exclusive use of residents.[87]

The Sellerses ensured that it would remain a place of large homes separated by ample green space. The Boston paper reported the developers were creating a fund for open space that would prevent overdevelopment. "This fund can never be used" to pay for "water and gas or to reduce taxes, but must ever remain as a guarantee, the property of the landholders, to ensure the preservation and maintenance of the public property as first planned and laid out."[88] In this separation of land uses and restrictions on growth, Ridley's planning and governance employed what would become key tools of exclusive suburban development for the affluent residents of American metropolises.[89]

At the same time that he invested in Ridley Park, William began another suburban project for a different group of people. The company town he developed at Edge Moor was part of a larger set of company towns in the region, most of which are not well documented by historians. In Philadelphia, major manufacturers such as the Disston Saw Works, the Dobson textile mills, and chemical makers Powers & Weightman built working-class suburbs-in-the-city.[90] Outside of cities, others created more isolated towns.

Manufacturing was a central driver of suburbanization. Nationally, mill towns and company towns dispersed production and working-class residence to a greater degree than most historians of elite and middle-class suburbanization have recognized.[91] For prominent manufacturers like the

Sellerses, company towns were also an instrument of competition with other national and international firms.

Like other industrialists who built housing and facilities for their workers, the Sellerses took a keen interest in British and American company towns, part of a larger transatlantic exchange of ideas about working-class housing.[92] On his 1884 trip to Europe, Coleman visited textile manufacturer Sir Titus Salt's town of Saltaire outside Leeds. "We are in the habit of speaking of the Town of Pullman [Illinois] as the one place in the world that is to be commended as a great private undertaking to build a city," he wrote to Coleman Jr. "Here at hand and now growing old," thirty years after its founding, "is a greater one I think." It had "fine stores . . . and very fine public buildings with costly carvings and much statuary," a hospital, bathhouses, a large park on the river, allotment gardens, and an "institute" for education and recreation, with a library, science lab, gymnasium, and concert hall. Coleman called the workers' accommodations "the best houses for that purpose I have seen in all England . . . nice two story stone houses."[93]

Coleman's praise for Saltaire suggests he favored a relatively generous version of the company town. Historians' interpretations of these ventures range from condemnation of their gross curtailment of social freedoms, as at Pullman, to celebration of benevolent paternalism, as in historian Harry Silcox's account of Disston.[94] With three thousand employees living there, Saltaire achieved a scale and boasted amenities far beyond those of Edge Moor and other company towns in America. Yet even tiny Edge Moor village had social programs and something of a civic life.

For the Sellerses, like other industrialists, the ultimate ends of these towns lay on the shop floor. In one of his many articles, Coleman stressed the importance of "good living accommodations for the working class," remarking, "The comfort and health of the workmen has much to do with the quality of the work they turn out and its cheapness."[95] Manufacturers' control of their workers' homes and social institutions promised to reduce turnover and the chances of labor unrest, making for a more stable and productive workforce.

The company town at Edge Moor and its surroundings illustrate the class dynamics of late nineteenth-century suburbanization in microcosm. At the same time that they established the company, William and his first wife, Mary, developed the estate of Clifton, their country house, on the first hilltop overlooking the marshy area of Edge Moor (literally on the cliff at the edge of the moor). They improved the site with tall English boxwood trees and stone retaining walls according to the fashion of picturesque British landscape gardening, much like Copeland's designs at Ridley, and possibly with his help.[96] It was very much a manor house, overlooking the mill in the stereotypical fashion of the industrialist who peers down upon the factory, keeping an eye on his workers. They built the workers' town in the image of a quaint village, complementing the adjacent pastures of the farm William established between his factory and house.

Clifton and Edge Moor were part of a larger pattern on the northern edge of Wilmington. Dotting the hills there were the estates of the city's affluent industrialists, including the neighboring mansion of Rockwood, owned by the Bringhursts. Their daughter Edith Ferris Bringhurst would marry Alexander Sellers, the son of William and his second wife, Amélie. Beginning in the 1880s, the Sellerses and Bringhursts joined their cousin William P. Bancroft in spearheading greater Wilmington's parks movement.[97] They created a string of parks buffering their estates from their working-class neighbors, while their properties in the hills escaped most of the air pollution from industry, thanks to prevailing winds.

To the east of Clifton, the Sellerses assembled virtually all the property between the house and the ironworks, which would accommodate their business and hobbies. In this low-lying area, William embarked on a second career as a gentleman farmer, breeding horses and cattle. His farmhands consumed much of their time planting and maintaining hedges and fences. They took special pains to accommodate the rail lines that cut across this pastureland to serve his factory, with landscaping to screen and separate the industrial and pastoral features of the area.[98] William joined with other nearby manufacturers, including the Du Ponts, to incorporate the Cherry Island Marsh Co., a land improvement association that assessed fees on its members to maintain the banks, canals, roads, and drains of the riverfront marshes where they located their factories.[99]

The Sellerses built the Edge Moor workers' housing at the same time as their factory, but it did not house the entire workforce. It presumably featured in William's pitch as he recruited labor from England. He kept it at the scale of a village, not a town or city like Pullman or Saltaire. Its inhabitants would come from the ranks of Edge Moor's longer-serving, usually higher-skilled workers, and only those who were married.[100] The firm's big contracts, and consequently the size of its workforce, fluctuated with business and real estate cycles. To secure a sufficient supply of additional labor at this somewhat isolated site, William incorporated the River Front Railroad, a spur off the Philadelphia, Wilmington & Baltimore line that brought men from Wilmington each morning.

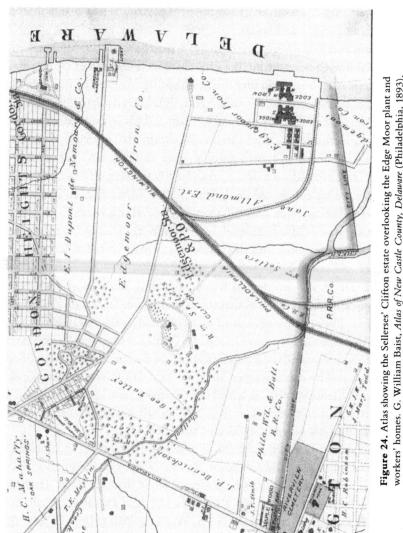

Figure 24. Atlas showing the Sellerses' Clifton estate overlooking the Edge Moor plant and workers' homes. G. William Baist, *Atlas of New Castle County, Delaware* (Philadelphia, 1893), plate 2. Library Company of Philadelphia.

Edge Moor village was a self-contained community, at least in the eyes of the Sellerses and their class. As historian Thomas Scharf related in 1888, married men had "the privilege of renting, at low rates, excellent cottages." The firm "also erected a convenient and commodious school-house, in which they maintain a school for the benefit of the children of operatives residing there."[101] Cinders from the factory's power house paved the few streets, and the villagers cultivated small gardens behind their homes. By the early twentieth century, the company sponsored an athletic association for employees, who played on the village's baseball diamond, tennis courts, and golf course. They also fished in the river nearby.[102]

The Sellerses made residents responsible for maintaining the place. The night watchman at the plant lit its streetlamps. Another man trimmed the grass along the roads and paths. Following a common pattern in American labor history, the least desirable job fell to an African American employee who cleaned the outhouse pits. In 1871, the roughly eighty inhabitants organized a Sunday school in the gatehouse of the ironworks. By 1876, residents were using the schoolhouse for religious services, but in 1885 William transferred a plot of ground to the Edgemoor Methodist Episcopal Church for the nominal fee of five dollars. The residents built a sanctuary the following year. This would be the only piece of property at Edge Moor that the Sellerses and their company did not own themselves, which in the twentieth century aided the removal of the village.[103]

William's interest in working-class housing was not limited to his employees. In 1869, about the same time he built Edge Moor, he joined fellow Social Science Association members in forming the Beneficent Building Association in Philadelphia.[104] This private experiment in social reform was based on the group's ability to marshal capital for real estate investment. They aimed to develop a model of slum upgrading for the public sector to emulate.

In particular, William and his colleagues sought to demonstrate how "scientific" intercession could clean up the slums while paying for itself with the rents they collected. They chose as their target the city's most notorious alley, Alaska Street in the slum of Moyamensing just south of the downtown. Despite the pleas of elite reformers and exposés by the newspapers, the city government had failed to do anything to change the area's fetid environment and inhabitants. The Beneficent Building Association took up subscriptions to purchase and demolish Alaska Street's old houses and erect new ones set farther back. They would widen the street to introduce more sanitary conditions of air and light, and rent to carefully chosen tenants who would serve as a positive influence for their neighbors.

By the centennial, they had sunk more than $56,000 into a dozen properties around Alaska Street, earning an annual return of about 3 percent. In their view, this proved the viability of private reform through the real estate market. It offered what today is called a market model of urban redevelopment.

The Sellerses' real estate ventures in the inner city and suburbs were different parts of the same project of metropolitan development. "Just as the political economy of outward growth sought to plant the seeds of domestic virtue on the city's suburban extremities," argues Andrew Heath, the Beneficent Building Association's "improvers sought to do the same on its inner frontier."[105] They claimed in their annual report for 1871 that "the physical reformation of the street, in its houses, and yards, and gutters" formed the "basis and support for the higher instrumentalities, moral and spiritual," of the residents of its "decayed and squalid tenements."[106] This stance would echo in the tenement reform movement that culminated in professional city planning a generation later.

The Beneficent Building Association was the real estate development arm of a larger complex of social reform institutions in which the Sellerses participated, including the Social Science Association and the Western Home for Poor Children. On the edge of Powelton, the Western Home represented another side of the elite's efforts to "domesticate" the working classes and support more vulnerable members of society. It housed "poor white children under the age of twelve years," entrusted to its care by their guardians or "committed to their management by any of the judges" of the city or state. John Sellers Jr., Clarence Clark, and Anthony Drexel served on its board of trustees. Its board of managers included the wives of other prominent businessmen.[107] While real estate was usually men's territory, contemporaries viewed direct care and influence of children's behavior among the important civic works of elite women.

The "moral environmentalism" of Philadelphia capitalists' critiques and solutions to urban problems was typical of elite urban reformers, and it proved a logical outgrowth of their evolving visions of a moral economy.[108] As Lippincott's magazine claimed in promoting Ridley Park, "The finest of old cities are the accretions of accident," and "after the capital becomes wealthy" an urban reformer "has to be sent for to rearrange the ill-judged furniture and reupholster the confused ornaments."[109] This rationale described the work of the Beneficent Building Association, Fairmount Park, and other urban and suburban development projects in addition to Ridley.

Beyond the poetic vision of the landscape architect or real estate boosters, the engineering and social research supported by the Sellerses built a scientific case that allowed them to claim certain authority to promote their plans. To be sure, their power was limited, and increasingly so as machine politicians pushed them out of positions of authority in the Park Commission, Union League, and other institutions. Cases like Edge Moor village and Alaska Street were small compared to other urban projects. However, through private development and reform movements and institutions, as well as public authorities they controlled for a time, the Sellerses and their allies exercised uneven, but sometimes substantial, influence over the region's environment in the late nineteenth century. In their collective investments and experiments, they advanced the piecemeal yet networked process of reorganizing the land-use patterns and social geography of the metropolis.

Electrifying the Metropolis

Some of the systems the Sellerses designed and built had especially pervasive impacts on metropolitan development. Coleman's work with electricity late in life was perhaps the best example of this. Philadelphia inventors are rarely mentioned in the history of electricity after the era of Ben Franklin. But for a moment in the 1880s and '90s it appeared that they would hold a central place in the electrification of manufacturing and of cities.[110] Perhaps no engineer reflected this promise more than Coleman, whose work in Philadelphia and at Niagara Falls served capitalists' plans to build new industrial sectors and the institutions, infrastructure, and urban forms that the second industrial revolution both depended upon and enabled.

The Sellerses pursued institutional strategies to strengthen Philadelphia as an early center of electrical engineering and allied industry. In 1882, Coleman, his son Horace, and other Sellers & Co. employees participated in forming a special department at the Franklin Institute for electrical research and development.[111] Two years later, they mounted an international electrical exhibition, for which they obtained funds from Congress to help underwrite a conference, on-site experiments, and the publication of their results.[112] The Pennsylvania Railroad donated a large lot next to its West Philadelphia terminal, and after the fair it agreed to keep the exhibit hall open for a "Novelties Exhibition" to popularize electrical technologies. However, attendance at the latter exhibit was low.[113]

In 1889, a group of New York investors approached Coleman about a plan to harness power from the waters of Niagara Falls to build a massive manufacturing center. They asked him to submit a report on the generation and transmission of power via channels that would siphon water from the falls to factories below. This would create a cluster of manufacturing comparable to the largest and densest mill towns of New England. The next year, these capitalists established the Niagara Falls Power Co. and the Cataract Construction Co. Their mission, as Coleman later recounted, was "to execute the work, to improve the lands owned or controlled by those interested and to furnish transportation facilities to a large industrial district, where a uniform water power, without fear of low water or freshets [floods], would be obtained."[114]

However, Cataract Co. president Edward Adams soon decided to abandon this plan for the ambitious and innovative project to generate and transmit electricity over a much broader area of western New York. He hired Coleman as its chief consulting engineer. Cataract paid him $10,000 per year ($244,000 in 2010) to oversee construction, subcontracting, and a testing laboratory built at the falls. Horace assisted in the project, manning Cataract's Philadelphia office while his father traveled to New York and overseas. Coleman toured European power plants and corresponded with pioneers of electrical engineering such as Thomas Edison and Nikola Tesla.[115]

This was the culmination of Coleman's lifelong experience with electrical technology, which presumably started with the collection of his grandfather Peale. Some of his earliest manufacturing included production of telegraph wire in Cincinnati. Prior to leaving Sellers & Co., he oversaw the installation of electric motors in heavy metalworking production. He also enjoyed a prominent place among networks and institutions of American and European engineering, which like the Franklin Institute were consumed with the flowering of electrical technology in the 1880s and '90s.

The centerpiece of Coleman's research and development program was a carefully crafted institution, the Niagara Falls International Commission. Writing to Edward Adams in 1889, he warned, "I hold it is unwise to depend upon any one or more persons for the technical knowledge required to fully develop the scheme," especially given the untried nature of the venture.[116] He knew that each inventor would simply push his own methods. Instead, by composing a commission of leading authorities on the generation and transmission of electrical power, the project could develop the most profitable system for the companies involved, not for any individual engineer. Coleman assembled a group of primarily academics, as opposed to entrepreneurs trying to sell their patents.

In addition to Coleman, the commission included eminent European engineers. They brought the company good publicity and lent the project an air of public and international significance, promoting confidence in a novel and challenging project.[117] From Britain he recruited Lord Kelvin, a professor of natural philosophy at Glasgow who had helped develop the science required to make the transatlantic telegraph, and Professor William Unwin, dean of London's South Kensington Technical School. Physics professor Éleuthère Mascart came from Paris, while Colonel Théodore Turrettini hailed from Geneva, where he was the city's mayor and engineer of its advanced electrical plant. With these colleagues' help, Coleman cultivated the project's public image among the world's scientific community, mainly through lectures and journal articles.

Out of their collective research and discussions, Coleman recommended a set of leading American manufacturers to build the plant. Dynamos from the Westinghouse Electric and Manufacturing Co., made with the help of Bethlehem Steel, were to generate alternating current, which would be transmitted via wires to factories and utilities. Sellers & Co. supplied testing devices and a massive crane, while Pratt & Whitney and I. P. Morris collaborated on other machinery for the Niagara power plant. This would be the first large-scale implementation of alternating current power generation and transmission, a major event in the history of engineering.

Alternating current helped foster a new generation of manufacturers associated with the second industrial revolution, especially in the electrical, steel, chemical, and petroleum sectors. As Coleman later wrote, "The advantages offered by [the] elastic electric system at Niagara Falls attracted industries that employed few hands, needed little machinery, but required enormous amounts of power." With wires carrying that power to factories "at a lower cost than steam and without an investment in engines and dynamos" at each plant, electricity helped make new forms of production profitable.[118]

Niagara attracted the big industry its New York investors initially envisioned, creating a classically second industrial revolution cluster virtually overnight. Firms concentrating around its central power plant included the Pittsburgh Reduction Co., "engaged in the extraction of the metal aluminum from its ores." The neighboring Carborundum Co. used "a process in which heat energy alone is needed to produce a new mineral next in hardness below the diamond." The huge Union Carbide Co. also located there, as did other electrochemical concerns forging the industrial materials of the future. In a move that limited the market for one of Sellers & Co.'s core product lines but increased demand for Edge Moor's boilers, big factories around the country "recognized the economy of generating electricity by steam and transmitting it to the various machines . . . thus dispensing with long lines of shafting."[119]

The Niagara project also heralded a new era for urbanization. Coleman reported in 1899, "electricity has been furnished to the lighting station" of the city of Niagara Falls, and "a direct current is being delivered to the trolley lines of Niagara Falls and Buffalo," twenty-two miles away.[120] The general public of western New York's principal urban center, and later the inhabitants of New York City, would enjoy the energy provided by Niagara's hydroelectric infrastructure. Alternating current would provide transportation, light, and ultimately power to drive the household appliances of the twentieth century. Within a generation, electrical lines ran down virtually every city street and through most buildings in urban America. Electrification of cities and later rural regions fundamentally altered people's experience of time, space, work, home, and leisure.

When he was not busy with electricity, Coleman spent much of the 1890s researching other technologies that would transform cities and material life in the coming century. With Professor Unwin and eminent electrical engineer Elihu Thomson, he corresponded about petroleum engines and the emulsion and refining of fuels for heating, manufacturing power, and soon motor vehicles.[121] With other Philadelphia capitalists, he debated the prospect of air travel, which Coleman believed possible. In an 1892 exchange with a skeptic, he cited the success of submarines in underwater navigation and "experiments now being made . . . as to the flight of birds." He speculated that the military advantages "when explosives can be dropped on to any city" might spur development of peaceful air navigation, as well.[122] Though he would not live to see it, Coleman could imagine subsequent technological and transportation revolutions, and their military backers, that would again remake cities and the relationships between them.

In their own lifetimes, during the late nineteenth century, Coleman and William helped mold other enduring processes and patterns of urban planning and growth. Second industrial revolution technologies would continue to define cities' material environment. Parks, suburban development, and metropolitan infrastructure would increasingly separate zones of work, home, and leisure, delineating the region's geography of industry and class to the present day. While their scope of work, from machine design to slum upgrading, would be divided between various professions in the twentieth century, the strategies and theories behind late nineteenth-century industrialists' real estate and social reform interventions would be repeated in later public and private planning.

The Sellerses' diverse activities illustrate how elite industrialists cultivated a science of city building and sought to apply it in developing and redeveloping a cross section of the metropolitan environment. They show how Americans with capital and influence, if not control of city government, could shape urban regions through civic institutions and property development, albeit unevenly of course. They also reveal how social and natural science contributed to a critique of industry in the urban environment that served their effort to reorganize the geography of production on a large scale. These interventions constituted the spatial side of economic development, as real estate development became an increasingly large part of the fast-growing city's economy. This was not necessarily a good trend for the economic development of manufacturing cities like Philadelphia.

CHAPTER 7

✗

ROOTS OF DECLINE

In April 1900, William and John Jr., both in their mid-seventies, sold the Edge Moor Bridge Works to the newly formed American Bridge Co. J. P. Morgan assembled the new corporation in a classic Morgan consolidation to reorder competition and profit in the industry. He merged Edge Moor Bridge with twenty-seven other firms, including Carnegie's Keystone Bridge. Morgan's group paid the Sellerses $321,247 (over $8.6 million in 2010) for their company and half of the right-of-way of the River Front Railway that served the factory.[1] The Sellerses had always struggled to make a profit in this business, and elected to sell rather than compete against the new concern with its large capital and networks of production that extended across the country. American Bridge kept the plant open another quarter century. In 1926 it closed the facility in Delaware as it consolidated operations around its headquarters outside Pittsburgh.[2]

William and Coleman's later years and the careers of their children reveal late nineteenth- and early twentieth-century changes that positioned metropolitan Philadelphia for industrial decline. The sale of Edge Moor Bridge demonstrates one of the primary ways through which power over regional economies shifted from local elites to national capitalists, usually bankers in New York. From the 1880s, William lived in relative isolation with his family at Clifton, another signal of the Sellerses' weakening networks and waning authority even in their own region.

It is unreasonable to expect a single family of engineers and industrialists to sustain its influence over the city and its economic development across so many periods of economic history. In this light, it is remarkable that the Sellerses did for so long. In the twentieth century, however, national capitalism would eclipse their mode of family capitalism.

Perhaps no one more than Coleman's son Horace pivoted so dramatically in response to these changes. Trained as a mechanical engineer, he abandoned his father's consulting practice in the mid-1890s. He founded an architectural firm in which he designed suburban homes and specialized in the restoration of such monuments as Independence Hall and George Washington's headquarters at Valley Forge. A member of the Sons of the Revolution, the Merion Cricket Club, and the Numismatic and Antiquarian Society of Philadelphia, he helped the region's elite reconnect with their past. He made signal contributions to redefining the city's national image, mainly in terms of its Revolutionary-era political history. This identity would grow in importance over time, especially when the city's industries crashed in the late twentieth century.

In many ways, the first half of the twentieth century remained a prosperous time for manufacturing in Philadelphia. Production and employment grew, except during the Great Depression. However, local capitalists' loss of power and the erosion of key sectors before the 1930s would have devastating consequences for the region following a momentary boom during World War II. In the second half of the century, the city and the region would lose nearly their entire factory economy. Between 1947, when nearly half of all jobs were in manufacturing, and the end of the century, more than 85 percent of those jobs disappeared.[3] This was a national story, an experience shared with cities from Baltimore up to Maine and west to St. Louis and the Great Lakes.

Since at least the 1970s, social scientists have asked why the industrial cities of the American Rust Belt fell so hard and so fast. The standard answers focus on postwar shifts in political economy, capital mobility, and cheaper labor in other places. Yet the roots of Rust Belt cities' decline are more complex and often deeper. A few historians have argued that changes in the structure of the American economy that took place around the turn of the twentieth century created conditions that enabled or even drove later plant closings and job losses.[4]

The Sellerses illuminate particular elements of this earlier transformation, showing how industrial capitalists responded to their region's declining power within the new national capitalism of the late nineteenth and early twentieth century. Their companies reflect several common pathways taken by family firms in old or restructuring sectors, including those of the first industrial revolution. Their careers illustrate larger trends in the region's complex of institutions and networks of capitalists. Their experiences reveal key dimensions in the reorganization of Philadelphia's economic geography that contributed to its deindustrialization in spatial terms, as well.

The forces undermining the region's competitive position from the end of the nineteenth century onward were largely beyond the control of local capitalists. Most telling were industrialists' inability or failure to position their firms and institutions competitively within the shifting global economy. Although Philadelphia remained an important manufacturing center, finance increasingly drove national economic development and defined cities' place in the world. As New York financiers created powerful corporations across major sectors of manufacturing, local firms sold out or clung to a form of proprietary capitalism that struggled to compete against the vast resources amassed by their national competitors.

Beginning around 1890, the region's manufacturers disinvested in the institutional complex that had supported industrial development for decades. By the 1930s, the Franklin Institute became a museum rather than an active engine of innovation. Penn lost many of its links to the regional economy, too. The choices behind these shifts were quite rational, responding to and reinforcing local industry's eroding power nationally. But they also limited local businessmen's involvement and investment in the region's broader development.

From the late nineteenth century, urban capitalists sought to deindustrialize key parts of the city. Politicians, planners, developers, and other interests mobilized to remove industry from increasingly large swaths of the city and region. Building upon the critiques of industry advanced by the parks and public health movements, in the 1890s Philadelphians set about pushing factories out of the central city practically altogether. Members of the Republican machine, who effectively defeated their reformist challengers, pushed through projects that compromised the viability of industry in Bush Hill and other districts long before their firms were obsolete. The machine's mode of economic development constituted little more than real estate development. This created jobs and built a vast metropolis, but it did not foster the innovation and human capital development through which earlier generations made Philadelphia an international center of industry. William Sellers resisted some of these changes toward the end of his life.

Deindustrializing the Urban Environment

When William died from kidney failure in January 1905, city planners and politicians had been scheming already for years to demolish his Philadelphia home on Logan Square. They dreamed of a grand civic boulevard between City Hall and the entrance to Fairmount Park. On his block, a new court building would later rise, designed by the chief architect of his enemies in the Republican machine. In the decade and a half before his passing, William had desperately opposed the forces seeking to restructure his neighborhood and especially adjacent Bush Hill. Less than a quarter century after his death, on the eve of the Great Depression, his family's tool works would be the only major machine builder left there. Baldwin, Bement, and many smaller concerns would be gone, the area's industries replaced to a great extent by museums and civic monuments of a new era.

The story of how this happened illustrates a little-appreciated force in the industrial restructuring of American cities. While the economic dimensions of this process are well understood, historians have explored its spatial dynamics much less. Around the turn of the twentieth century, urban planners and their allies in corporate and government circles dreamed up a "city beautiful." It would replace some of the less handsome but economically vital parts of Chicago, Philadelphia, St. Louis, and other centers.[5]

The urban reform movements the Sellerses helped lead in the post–Civil War era advanced their critiques of industry in ways that William clearly came to regret. Eliminating railroads from city streets became a major objective of Progressive urbanists, an important part of their environmental reform and public safety campaigns. Whereas boosters of the 1840s and '50s had celebrated railroad construction, as early as 1866 state legislators from Philadelphia pushed through a statute designed to remove the City Railroad from Broad Street.[6]

Deindustrializing urban landscapes would become a large part of early professional planners' attempts to reenvision central cities around the turn of the twentieth century. The plans for grand boulevards and civic spaces put forth by Daniel Burnham and fellow City Beautiful designers sought to erase large expanses of rail yards, factories, and other industrial facilities. Their parkways facilitated the growth of early automobile suburbs. The rhetoric and results of their projects helped early professional planning redefine the place of industry in American cities. The public health concerns behind eliminating grade crossings were real. But in the minds of many civic leaders and architects shaping the physical environment of the metropolis, factories and rail lines became nuisance land uses more generally, impediments to building a world-class city.

As the Sellerses and their allies lost control of the city's planning and growth to the construction, utility, and real estate magnates allied with the Republican machine, the deindustrialization of their central city factory districts became possible. According to muckraker Lincoln Steffens, who in 1903 labeled the city "corrupt and contented," the contractor bosses and utility monopolists of the machine made "public franchises, public works, and public contracts . . . the principal branches of the business" of government.[7] They made real estate development the heart of urban economic development.

In the process, they helped undermine the viability of manufacturing in the inner city. In the 1890s, the city government forced the Reading Railroad to bury its line along Pennsylvania Avenue in a "subway" that restricted plant expansion in Bush Hill. The subsequent development of the Fairmount Parkway demolished the western portion of the district, including the two plants of the Bement tool company. The boulevard remade this portion of the city as part arts center and part highway to the suburbs, accelerating manufacturers' departure for new suburban plants. During the parkway's construction between 1906 and 1928, the Baldwin Locomotive Works built and transferred operations to a massive site in Delaware County, removing from the city its largest employer.

The sinking of the Reading line along Pennsylvania Avenue between 1894 and 1902 was part of broader efforts by public authorities to rid the central city of railroad tracks running at grade. For the Reading, it represented a tradeoff, as its managers agreed to the plan in return for permission to erect a new passenger terminal on Market Street, at the center of the downtown core of department stores.[8] Although the railroad sued the city over elements of the plan, much of this amounted to posturing for a better deal. William Sellers was the project's main public critic.[9]

To William, the newspapers' announcement of the project in January 1889 came as a surprise. He was a man accustomed to knowing about deals before they were cut, but late in his life other Philadelphia capitalists increasingly cut him out of the circles of power. "The arrangement reported in the Press must have been reached without consultation with the manufacturing interests on Penna. Ave.," he wrote fellow political reformer Henry Charles Lea. Not even his old colleague Joseph Wilson, the project's civil engineer, had informed him of the plan. "I cannot believe," William wrote Lea, "that any manufacturer upon this Avenue would assent to such a curtailment of their railroad connections."[10] But he would be virtually alone in his opposition.

In galvanizing support for this project, politicians and the press campaigned against the evils of rail lines and manufacturing in the central city. Mayor Edwin Stuart asserted that the tracks along Pennsylvania Avenue "have ever been a menace to life and a most serious obstruction not only to the public and business generally, but have also acted as a most potent factor against the proper growth, development and improvement of that entire section of the city."[11] The machine-allied *Philadelphia Times* speculated, "It may be that in the course of time a more profitable use can be found for the property upon Bush Hill than that of manufacturing. The abolition of the surface tracks . . . at least will remove the obstacles that now make any improvement impossible."[12]

The *Philadelphia Evening Bulletin* reasoned that the city's plan to split the cost with the financially troubled railroad was justified by its duty to regulate urban growth. "The city's responsibility for the present unfortunate state of affairs . . . is not heavy," it editorialized. However, "since it made no effort years ago to avoid grade crossings in laying out streets in those parts where the Reading Road already had run its tracks while the surrounding country was not yet built up, it cannot altogether escape the blame for what, as is now plainly seen, was a mistake." It celebrated "the reclamation of what should be ultimately one of the finest sections of the city, and not as it is now, a half-blighted region."[13]

The railroad's response cited "most serious loss and inconvenience" and its "legal right . . . to maintain the existing status."[14] William editorialized in the role of rational engineer, suggesting the project's cost estimate was far too low and pointing out that a layer of quicksand underneath Pennsylvania Avenue threatened the viability of the plan.[15] But he did not win the argument with this appeal to "scientific" urbanism. He was the only person to testify against the proposal in city council hearings on the subject, and the newspapers dismissed his objections.[16] The *Times* ridiculed him, and noted, "No great public improvement was ever carried to successful completion that did not affect or threaten to affect unfavorably some private interests and call forth objection from those who feared individual disaster."[17]

William's isolated opposition to the Pennsylvania Avenue Subway signaled his diminished influence over public affairs in the city. Beyond his own isolation, elite Republican industrialists like the Sellerses no longer controlled the means—including government, civic institutions, and the press—to define the public good. Their relationship with their

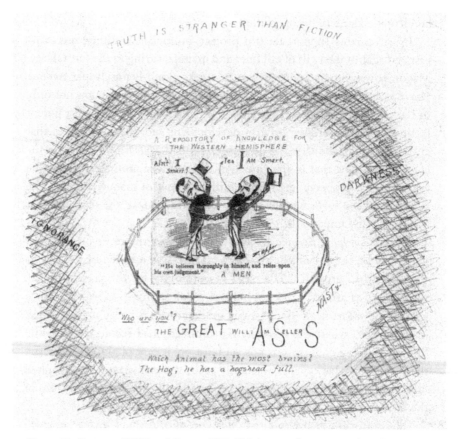

Figure 25. Cartoon of William Sellers, c. 1900. This homemade cartoon in the collection
of Baldwin Locomotive director William L. Austin indicates something of the animosity
that developed between Baldwin's partners and Sellers by the turn of the twentieth century.
Two caricatures of "the GREAT williAm SellerS," one "hot"-headed and the other "light"-
headed, congratulate each other on how smart they are, while surrounded by a cloud of
"ignorance" and "darkness." Austin used a clipping from cartoonist Thomas Nast and
elaborated his own "NASTy" adaptation (#Acc 179, from William L. Austin Collection,
Hagley Museum and Library). Hagley Museum and Library.

most important customer, the powerful Pennsylvania Railroad, soured
in the 1880s, as the railroad's new president, George Roberts, allied with
the machine.[18] In private correspondence between husband and wife,
William called him "wishy washy" in awarding contracts to other firms
that engaged in "bribery and corruption," while Amélie replied that Roberts was "a very narrow-minded man."[19] The Sellerses' alliance with the
Baldwin Locomotive Works' partners also deteriorated. In his autobiography, Samuel Vauclain portrayed William as a man who had, by the 1890s,

lost touch with the times and who vainly opposed the modernization of industry and the city.[20]

Building upon the momentum of the Pennsylvania Avenue Subway, the subsequent development of the Fairmount Parkway realized much of the vision for a new, deindustrialized Bush Hill and central Philadelphia.[21] In 1891, supporters submitted to the city council a petition signed by "500 influential citizens" who claimed to own more than two-thirds of the city's taxable property. They called for "the opening of a suitable and handsome boulevard" to the park "from that portion of the city most frequented by visitors and by business men coming from other parts of the country."[22] Machine politician James Beck, responding to the vastly popular neoclassical fairgrounds at the world's fair in Chicago two years later, expounded on the importance of this project: "A city, which is subjected under modern conditions to keenest competition with other cities, must be beautiful in order to gain business and attract the strangers within its gates." Yet factory districts still clogged large portions of the central city. This shortcoming and the lack of civic pride it occasioned, he wrote, "prevented Philadelphia from remaining in the beginning of the second century of our republic as she was at the commencement of the first—the metropolis of the western hemisphere."[23]

In 1903, the city council approved a plan for the parkway that would turn Logan Square into a large traffic circle around a central obelisk. It depicted a ring of mansions in place of the four-story town houses of William Sellers and Rudolph "the Dutch cleanser" Blankenburg, whose reformist mayoral administration (1911–15) would be the only interlude in more than sixty years of machine rule.[24] The plan also showed the parkway cutting through Bement's plants and the western part of Bush Hill, where hundreds of people lived and where textile mills and smaller metalworkers employed thousands of men and women.

The parkway's boosters elaborated arguments that would echo in urban renewal later in the century. The city's Department of Public Works predicted the plan "would greatly relieve traffic conditions in the center of the city, . . . would enhance the value of property which is now stationary or declining, would aid in the elimination of slum districts, and would add a feature of great distinction and usefulness to the city."[25] In place of dense blocks of factories and worker housing, the architects and the press explained, Philadelphia's downtown needed to grow in new ways, with wide streets, grand vistas, and cultural attractions. While planners erased the mansions from the plan, they reconceived the boulevard as an avenue lined by cultural institutions. Its crowning jewel would

be the Philadelphia Museum of Art above the old waterworks at the edge
of the park. Architect Horace Trumbauer, the same man who designed
the library to replace the homes of William Sellers and his neighbors,
planned the museum's three wings to house the collections of utility
monopolists Peter Widener and William Elkins and their lawyer John
Johnson.

Like the park commission in the 1860s, the parkway's promoters used
the state legislature to give public institutions new powers over planning
and urban development. A state law passed in 1907 permitted cities to
purchase and condemn land two hundred feet back from the edges of
parks, creating effective eminent domain and zoning powers in an era
before zoning codes.[26] Demolition began that year. In 1912, machine-
elected judges appointed Widener's banker, E. T. Stotesbury, to head the
machine-dominated park commission. The commission laid unofficial
claim to the boulevard, battling Mayor Blankenburg, who refused to direct municipal funds to the project during his term. In 1916, with the
election of a machine mayor, the city granted the park commission juris-
diction over the entire parkway, including Logan Square.[27] In November
of that year, the city council assigned a zoning commission to work with
the park commissioners, adding a new city planning tool for regulating
land uses in the area.[28]

In the end, the project removed far more than it built. Civic institutions and potential funders showed lackluster interest in development
along the boulevard, conditions exacerbated by the economy of World
War I. In response, Stotesbury's architects erased most of the buildings
planned along its edges.[29] The road cut a broad swath through the western portion of Bush Hill, taking out metalworking, textile, and garment
factories as well as shops, institutions including the Medico-Chirurgical
College and Hospital, and hundreds of homes.

In 1925, the city demolished Bement's main plant and completed the
court building where William Sellers's house once stood. A key reason
why the parkway easily wiped the Bement works off the map lay in the
fact that Bement had merged to form the Niles-Bement-Pond conglomerate in 1899. Headquartered in New York, N-B-P also had plants in
New Jersey, Connecticut, and the old Niles factory where Coleman had
worked in Cincinnati.[30] When the parkway's development coincided with
a downturn in demand for machine tools following World War I, the
choice to close its Philadelphia operations was, like American Bridge's
plant closing at Edge Moor, not a difficult one.

Figure 26. Aerial photograph of Bush Hill in 1925, with Baldwin's plant in the foreground and the new parkway beyond the factories that remained. *Baldwin Works*, #5164. August 1925, Aero Service Corp. Photo courtesy of Aerial Viewpoint, Inc., Spring, Texas.

The city estimated the ongoing project's total cost at $74 million in the mid-1920s ($920 million in 2010 dollars). However, the private development and rising property tax revenues intended to pay for the investment did not materialize. Instead, the parkway would serve as the initial link in the city's first major automobile highway from the core of office towers around city hall, including the Widener Building, to the northwestern suburbs, where Stotesbury, the Wideners, and other affluent Philadelphians resided. The parkway's planners helped establish a lasting pattern of metropolitan improvement through factory removal and the construction of highway connections between downtown and the suburbs, which would become main features of the urban renewal programs of the midcentury.

Both despite and because of the new tools and powers developed by urban planners, the place of industry in the city became increasingly unclear. In 1918, the region's biggest building trades journal lauded the city's first proposed zoning ordinance, but questioned its treatment of industrial land uses. "The form which factory districts, as distinguished from business districts, should assume . . . is one which has as yet not been satisfactorily answered." Planners described the existing geography of manufacturing districts, but had not sorted out whether manufacturing "should be confined in the heart of the city, dispersed to the suburbs, segregated in concentric zones, laid out along lines radiating from the center of the city, arranged in parallel districts bisecting the city, or discriminately scattered throughout."[31]

City government had seemingly not made room for factories and industrial districts in its conception of twentieth-century development. Moreover, while reform-minded citizens criticized the politicians' handling of the parkway contracts and negotiations with property owners in its path, there was little public outcry regarding the demolition of factories.[32] William Sellers's isolation in the early debate over the project, before his death, suggests that manufacturers did not ally to defend or otherwise represent their interests in the urban development that would remove their central city factories.

For heavy manufacturers, the future was in the suburbs. Like other large machine builders in North America, which were developing single-story factories on suburban sites where they could take advantage of recent innovations in materials flow technology and operations management, the Baldwin Locomotive Works would soon leave the inner city. On Saturday, June 30, 1928, Samuel Vauclain closed the doors to the old Broad Street factory forever, removing the region's largest employer and its tax revenues from the city to a new, six-hundred-acre plant in Delaware County.[33] Over the previous decade and a half, North Broad Street had become home to some fourteen major automobile assembly plants. With showrooms on the ground floor and production housed in the upper stories of these reinforced concrete skyscrapers, companies from Packard and Hupmobile to Fiat and Cadillac promised to buoy the local metalworking sector. However, all of these factories were branch plants, using parts produced in Detroit or Ohio and controlled by corporate headquarters in other cities, and their Philadelphia operations were short-lived.

Baldwin marketed its property to developers who would remake the area in the image of the modern metropolis. In 1926, the company commissioned a plan for its blocks on Broad Street with a theater flanked by office buildings.[34] Between 1939 and 1950, it sold off the property in large parcels, but buyers did not build the anticipated commercial, residential, or institutional facilities. Plans for a professional hockey arena and a public school baseball and football stadium came to naught.[35]

Instead, the area became home to warehouses and newspaper plants that, compared to the metalworkers they replaced, created few well-paying blue-collar jobs. Not coincidentally, the first major post–World War II race riots in Philadelphia occurred in Spring Garden in 1952. Working-class whites whose factory jobs were disappearing attacked Puerto Ricans who were moving into the neighborhood, attracted by its affordable housing and low-wage service jobs.

Bush Hill was home to two prominent firms in the mid-twentieth century, Electric Storage Battery and Smith Kline French (SKF) pharmaceuticals, though they both reflect the larger trajectory of Philadelphia manufacturing. The first firm, founded by a vice president of the machine's United Gas Improvement Co., supplied batteries to the machine's streetcar company, and is today the world's second-largest maker of acid batteries for autos and industrial use.[36] Renamed Exide, it moved most of its local production to the suburbs in 1958. It is now headquartered in Georgia, with factories across the non-union South. SKF is today part of GlaxoSmithKline, one of the world's biggest pharmaceutical companies, headquartered in the United Kingdom. It retains an office in Philadelphia, though it has moved from the edge of Bush Hill to a tax-free zone at the city's former navy yard.

Though just one of many converging forces reshaping industrial centers in the twentieth century, the demolition of factory districts and resulting loss of jobs played a key role in deindustrializing the economy, environment, and labor market of northeastern and midwestern cities.[37] City Beautiful planners and their backers established patterns that would be repeated by prominent postwar planners such as New York's Robert Moses and Boston's Ed Logue: factory removal in the central city; clearing of working-class neighborhoods to accommodate highways to the suburbs; and costly redevelopment schemes initiated by elite urban designers with real estate and contracting interests allied with politicians. Although City Beautiful improvements such as the Fairmount Parkway helped elevate Philadelphia's status as a cultural center, they also aided its decline as an industrial metropolis. They ultimately did little to reposition older manufacturing centers to compete in the changing landscape of industrial capitalism, though they did help reorient large areas toward a new, if much smaller, economy of arts, culture, and tourism.

Economic and Social Restructuring

As suggested by the work of Edge Moor Bridge and of Coleman at Niagara, Philadelphia engineers remained to a great extent on the leading edge of second industrial revolution technologies. But they did not, by and large, turn those innovations into profitable, lasting, local enterprises that made the region an economic center of the second industrial revolution. As national corporations like American Bridge and U.S. Steel came to dominate entire sectors, Philadelphia and its firms lost their centrality in the nation's economic geography. In the first half of the twentieth century, the Sellerses and other major manufacturers would dismantle their firms.

The major changes in industrial capitalism that began in the late nineteenth century largely explain how these manufacturers lost the national influence and international markets they enjoyed in the generation following the Civil War. Starting in the 1880s and accelerating after the Panic of 1893, Wall Street financiers spearheaded the consolidation of conglomerates that came to dominate key sectors of American industry, including steel, machinery, electrical equipment, and mass consumer goods. Philadelphia manufacturers found themselves in a weak position in this context, partly because of the limits of their city's financial institutions. They typically ended up like Millbourne Mills, as small, poorly capitalized competitors in sectors of large-scale enterprise, or like Edge Moor Bridge and Bement's tool works, as branch plants of companies headquartered in other cities, often Manhattan.[38] This removed power over Philadelphia's industrial development, including decisions about investment and plant closures, from the region and its residents.

William and Coleman Sellers bequeathed historically important firms but relatively limited future opportunities to their heirs. And most of those heirs did little to sustain the family businesses, arguably none of which were well positioned to compete in the twentieth century. William's eldest son, William Ferris Sellers, managed the Edge Moor boiler company after his father's death, while Coleman Sellers Jr. ran the machine works. But most of their siblings, cousins, and children collected dividends rather than working in the factory themselves. They tended not to reinvest profits in plant expansion and new product development. When Millbourne Mills, Edge Moor Iron, and Sellers & Co. became unprofitable, they sold them off.

The first Sellers firm to close was Millbourne, which filed for bankruptcy following John Jr.'s death in 1907. Five years earlier, the family had handed over the company's presidency to a man named Richard Dewees, though William and John Jr. continued to invest hundreds of thousands of dollars in new storage tanks, railroad sidings, and other plant improvements. But when the two brothers' heirs called for an audit in 1907, they found that the mill was worth $500,000 yet had some $750,000 in debt and apparently insufficient profits to cover this. In order to reclaim the $281,000 owed to the estate of John Sellers Jr. and $140,600 due William's estate, they immediately filed for bankruptcy.[39] Though Dewees disputed outside assessments of the company's distressed finances, Millbourne's assets were sold, its creditors paid off, and the mill torn down to make way for a Sears department store in 1926.

A similar turn of events brought about the demise of the Edge Moor boiler works, though it took longer to close. At the time of William and John Jr.'s deaths, the firm owed the brothers $594,837 and $297,419, respectively. According to a later company memo, "It was long recognized that the funds represented by the debt were required as capital in the business." In 1919, Edge Moor issued William's estate six thousand new shares "as an offset to the debt. . . . Being unable to make a similar arrangement with the Estate of John Sellers, Jr., the company . . . paid in cash its debt" to his heirs.[40]

Edge Moor's capacity to pay off John Jr.'s estate came from its substantial earnings during World War I. Its good times continued through 1926. This helped William Ferris Sellers acquire nearly 90 percent of the six thousand shares from his brothers and sisters. He lived at Clifton with his second wife (and distant cousin) Mary Bancroft, working at the firm down the hill until shortly before his death in 1933. Although his father's generation had never drawn dividends from Edge Moor, William paid himself and his siblings $6 per share between December 1919 and 1931, a total of $432,000 over twelve years (almost $5.4 million in 2010).[41] However, for the ten-year period ending in 1933, Edge Moor incurred losses close to $300,000.[42]

William Ferris Sellers died holding a 48 percent interest in the concern. Since his children did not work at the plant, the vast majority—if not all—of the company's stock was owned by family members not connected with its operation.[43] His heirs commissioned an appraisal to determine the value of their securities. Using the dismal five-year period from 1928 to 1933, the appraiser naturally returned with bad news. An audit produced in 1932 had found an "apparent lack of effective, executive leadership for a period of at least ten years, resulting in expensive operation and loss of prestige." This, combined with an "accrued obsolescence in equipment, methods and developmental work," led to a "failure to recognize sufficiently early the rapid developments in steam generating equipment" following World War I. This "rendered obsolete" Edge Moor's boilers, "resulting in loss of position in the industry due to inadequacy of products for the trade."[44] Together, the sad state of affairs at the plant and a glut of capacity in the steam boiler industry made the company's prospects dismal.

As at Millbourne, the recently appointed president of Edge Moor, W. E. S. Dyer, disputed these findings. His claims reflect the state of denial that constitutes one of the more perverse dimensions of the region's manufacturing decline. Citing the firm's "most excellent reputation through the United States and many foreign countries," in February 1935 Dyer assured the stockholders that he was "still enthusiastic about the ultimate outlook for the future success of this Company," which he believed was "in first-class shape to obtain a satisfactory share" of future demand in the industry.[45] In November, however, stockholders from the extended Sellers and Garrett families voted to liquidate Edge Moor Iron and distribute all assets among themselves.

Already in July, the Wilmington press had reported that the "cozy industrial hamlet" in the shadow of the plant would be demolished to make way for a large textile mill. The "last obstacle was removed" when trustees of the village church agreed to sell their property.[46] A group of Canadian and American investors led by Dyer purchased the boiler plant, and DuPont built a large paint factory on another part of the Sellerses' old landholdings.[47] Following World War II, the Condenser Service & Engineering Co. of Hoboken, New Jersey, purchased the boiler works. But in 1952 the firm was dissolved and the property sold to the Graver Tank Co. of Chicago, which by 1970 sold its machinery for scrap.[48]

As a giant in its industry, William Sellers & Co. had farther to fall than Millbourne or Edge Moor. People named Sellers continued to work at the Bush Hill company, which still enjoyed a prestigious reputation, but the family's commitment to machine building waned. Coleman Jr. ran the plant until his death in 1922. William's grandson William Sellers II served as its vice president from his Princeton graduation in 1923 until its sale in 1943.[49] William II had a master's degree in engineering, but he was apparently more interested in racing powerboats and flying airplanes than in machine tools. He raced to Bermuda "numerous times" and once took Amelia Earhart to dinner.[50]

Sellers & Co. continued to operate as a tightly held corporation, making planers, lathes, cranes, and steam injectors. However, the shop floor and drafting office did not sustain their track record of innovation.

The works' position in the machine tool industry declined as the market for its railroad equipment slowly withered. Sales soared during both world wars, but as at other leading metalworkers in Philadelphia, military contracts proved too variable to maintain the business. Electricity and diesel engines largely did away with demand for Sellers & Co.'s shafting and steam injectors (and Baldwin's steam locomotives). The firm missed an opportunity to develop product lines for bicycle and automobile makers, which became the province of Cincinnati tool builders.[51]

The district of Bush Hill around Sellers & Co. no longer helped sustain its resident firms through vibrant networks of innovation and production. In addition to Baldwin, Sellers's neighbors Asa Whitney and Hoopes & Townsend closed in the early twentieth century. The automakers on Broad Street began to close already in 1928, as the industry consolidated around Detroit.[52] As the anchors of the city's metalworking districts faded, folded, or sold out, smaller firms dependent upon subcontracting networks likewise disappeared.[53]

When his uncle William Ferris Sellers died in 1933, William II, his father Alexander Sellers, and cousin J. Sellers Bancroft Jr. looked into liquidating the firm. The tool works' financial condition remained "very strong," with assets of $1.13 million and liabilities totaling less than $4,000. Nonetheless, their appraiser concluded, "The future outlook for the William Sellers & Co. . . . at best looks only moderately favorable," as "there would seem to be little . . . demand for Producers Goods for some time." "The Company may be in a position," the appraiser argued, "that requires the designing and marketing of additional profitable products before operating profits could be shown."[54]

Losses reached nearly a quarter million dollars in 1932 and again in 1938. In September 1938 the Sellers family contracted with an outside consultant, Clarke-Harrison Inc., to take over the company's management. At their recommendation, the family sold some of the stocks, bonds, and real estate held by the firm. After an investment of $130,000 in plant and machinery upgrades, Sellers & Co. posted profits over the next two years, mainly from U.S. defense program contracts in the run-up to World War II.[55] With the value of their stock boosted by wartime business, in the late summer of 1942 the Sellers heirs set out to find a buyer for the tool works.

Other major metalworkers in Philadelphia reflected the same shifts and challenges experienced by the Sellers firms in the first half of the twentieth century. "During the 1920s," according to historian Albert Churella, "Baldwin's dividend policies prevented the company from accumulating the cash reserves necessary to finance a comprehensive diesel locomotive research and development program during the 1930s."[56] Some firms sought mergers, though never with much success. Morgan's group blocked Cramp's shipyard, the second-largest industrial employer in the region in 1900, from merging with Bethlehem Steel and a British shipbuilder, snatching Bethlehem for their U.S. Shipbuilding conglomerate. With virtually no orders from the navy after World War I, Cramp closed in July 1927.[57] Baldwin purchased other metalworkers in Philadelphia, San Francisco, and Illinois in 1929, aiming to diversify its product line.[58] But it slipped into receivership as the market for its products dried up in the Depression. After the momentary revival of World War II, it made its last locomotive in 1956.[59]

The depression came early in the region's chief manufacturing sectors. In 1921, the local press reported that more than 80,000 people were out of work in metalworking, textiles, and other industries.[60] The city lost almost 17,000 metalworking jobs between 1921 and 1927, and another 2,300 disappeared by 1930—altogether, an almost 24 percent decline during the decade, much of this among old railroad suppliers and shipbuilders.[61] Even before the crash of 1929, Philadelphia had lost its position as America's foremost center of heavy industry.

The entire U.S. heavy machinery sector declined after 1900 because of structural changes in the economy, including the rise of consumer goods manufacturing and the decline of railroads.[62] Philadelphia fared especially poorly. The region's industries were deeply rooted in first industrial revolution sectors. Its specialist and batch producers were poorly positioned to shift production to serve mass markets.[63] Their close proprietary control and the weakness of the city's financial sector relative to nearby New York combined to leave them on the selling end of corporate mergers.[64] Mid-twentieth-century shifts in the military-industrial complex, once the province of the navy and northeastern manufacturers, favored the aerospace industries of the South and West.[65] The Sellers firms embodied all of these challenges.

The pattern of family capitalism that characterized the Sellerses in the first half of the twentieth century exemplifies how many elite manufacturers across the Northeast and Midwest responded to declining competitive advantages in their firms and regions. Decreasing involvement in the work of family factories, reaping dividends rather than reinvesting in innovation and production, and selling out to non-local capitalists to salvage an inheritance were all logical responses to larger economic changes. These behaviors both reflected and accelerated these families' loss of control over their own regions' economic development.

Other manufacturing families followed similar paths, including the owners of the region's seventh-largest employer in 1900, the Disston Saw Works, and the much smaller Boekel scientific instrument firm, both old Sellers & Co. customers.[66] Bush Hill brewer Louis Bergdoll, the man principally responsible for bringing auto production to Broad Street in the 1910s, had wooed Packard, Fiat, and other makers largely at the encouragement of his two sons. However, the younger Bergdolls' interest extended only as far as racing cars. One brother won the last annual Fairmount Park Race in 1911, but neither son went into manufacturing to perpetuate what they had urged their father to initiate.[67]

At the beginning of 1943, the Sellerses accepted an offer of $1.1 million cash (almost $14 million in 2010) from a group of New York investors who merged William Sellers & Co. into the Consolidated Machine Tool Co. of Rochester and closed the Bush Hill plant. The group sold the property at 1600 Hamilton Street to a real estate speculator.[68] Together with the adjacent U.S. Mint, the site eventually became part of the Community College of Philadelphia, an institution of remedial education and job training as opposed to research and innovation. This result is suggestive of the broader decline of the region's institutions.

Institutional Restructuring

Philadelphia's nineteenth-century industrial prosperity was rooted as much in its evolving complex of institutions as in its leading firms. Many of the institutions built by the Sellerses and their colleagues survive to the present day. But at the end of the nineteenth century the networks that tied them to the region's factories began to disintegrate. This is one reason they lost much of their influence over urban and economic development. Beyond the growing power of the Republican machine, the weaknesses of local manufacturers themselves diminished the influence of allied institutions. Conversely, as Philadelphia firms and entire sectors struggled, the region's institutions proved unable to reposition or revive local industry.

Recognizing that the forces of economic development around them were changing, in January 1887 the Franklin Institute's board appointed William Sellers and several colleagues to a committee to "take into consideration the future work of the Institute, and to suggest plans or methods for extension." The institution's finances were stretched, members demanded a larger building, and their influence in their own region was waning.[69] But William's committee failed to engineer another transformation as it had in the 1860s. Instead, the institute's role as a catalyst of regional economic development eroded over the next generation. This was both a symptom and a contributing factor in the disintegration of the networks of firms and institutions that had driven urbanization and industrialization in previous generations.

The institute's lectures in 1887 still focused on the latest technical questions of metallurgy, electricity, public health, and city building. But its annual report lamented, "it is embarrassed in every direction . . . above all" for "want of funds."[70] The board seemed perplexed that more Philadelphians would not support its vital mission. In their 1891 annual report, they touted the institute's multiplier effects in the region's economy, claiming "every cent added to its income adds so much more to the actual wealth of the city through its manufacturing interests."[71] Yet it lost members in each of the next several years. Enrollment in its educational programs also declined.[72] Its annual report for 1893 decried the lack of pride and investment in an institution often copied by the leading engineering societies of Boston, New York, and other cities, and which served as a think tank, developer of human capital, and promoter of innovation.[73]

When William died in 1905, his long list of honorary pallbearers included eminent trustees of the University of Pennsylvania, the Union League, and the Franklin Institute.[74] Yet none of these men enjoyed the power they had wielded less than a generation earlier. Their institutions were not major players in national capitalism, and even their influence within the region had waned.

In the 1890s, William and other leaders of Philadelphia's post–Civil War engineering community disengaged or were pushed out of leadership in the institutions through which they had shaped science, technology, and manufacturing. In 1892, the Sellerses and Bements fought off an attempt to push William off the board of the Academy of Natural Sciences.[75] Two years later, as Franklin Institute president Joseph Wilson proposed the Pennsylvania Avenue Subway and directors of Midvale Steel were elected to the board of managers, William sent a letter declining renomination to the board.[76] William and Coleman's sway in national and international engineering institutions also seems to have weakened in this period, and they lacked strong connections on Wall Street just as these ties became increasingly important for industry.

William's successor at the Bush Hill tool works, Coleman Jr., would later serve as vice president of the Franklin Institute, but the organization's usefulness for his and other firms diminished. In 1898, the institute took its first steps "to extend its operations by the popularization of the arts and sciences," establishing departments of fine arts, geography and travel, geology, horticulture, household economics, law, and literature, among other subjects.[77] The next year's lecture series immediately reflected this change, including talks titled "City of the Sultan," "The Snake-Dancers of the United States," and "The Development of Gothic Architecture as Shown in the English Cathedrals."[78] In 1923, the institute closed its school. Its journal came to focus exclusively on mathematics, not engineering questions applied to industry and urbanization. In the 1930s, the board decided to use its endowment, including a large bequest from soap manufacturer Samuel Fels, to erect a museum of technology and memorial to Benjamin Franklin on the Fairmount Parkway. It became a place of schoolchildren's exhibits, no longer an institution with meaningful connections to university engineering departments, nor an engine of the region's industrial development.

There were valid reasons for these changes, beyond the institute's self-preservation. In the shifting landscape of national corporate capitalism, many of the incentives for regional collaboration disappeared, and allied institutions responded accordingly. No longer, for example, did the managers of Niles-Bement-Pond's Bush Hill plant share an interest in maintaining the region as the center of American tool building. Trade associations like William Sellers and Joseph Wharton's American Iron and Steel Association were subsumed by truly national interests such as U.S. Steel. The Union League became a social club of lawyers, bankers, and politicians. The Saturday Club failed to survive beyond William's generation. The Philadelphia Manufacturers' Club became best known for the golf course it developed in the 1920s.

As the urban reform movements the Sellers and their colleagues led in the post–Civil War decades were professionalized, the institutions and work of planning, engineering, and public health grew increasingly separate, bureaucratic, and sometimes national. Some civic projects of elite reformers became the province of government or of competing organizations. By the 1890s, the Beneficent Building Association had endured a quarter century, but it had done little to change inner city housing and was hardly a major player in a crowded landscape of organizations seeking to remake the immigrant slums.[79] In the early 1920s, the Wideners would consolidate most of the city's nongovernmental health and poor relief associations, from the Tuberculosis Society to the Children's Aid Society, in a single Social Service Building, signaling their dominance of private charity in this era.[80] The Social Science Association, once an important instrument of elite reformers, was eclipsed by the rise of increasingly compartmentalized academic disciplines and professions, each with its own national associations usually headquartered somewhere besides Philadelphia.

The University of Pennsylvania failed to develop a strong endowment in the early twentieth century, continuing to depend on benefactors from the region's declining heavy industry for annual support.[81] The Philadelphians who graduated from its Wharton School often managed their families' factories into the postwar period, but most of those firms produced few new patents or lines of business. The university ultimately transcended the limits of the regional economy by attracting national figures to its faculty and administration, and by gaining federal funding for engineering and medicine in the later twentieth century.

But Penn also illustrates the region's restricted ability to turn innovation into economic development. In perhaps the greatest story of lost opportunity, when electrical engineers at the university developed the world's first supercomputer, ENIAC, in 1946, their employer prohibited them from turning a profit on the invention, citing standards of academic integrity. Although the professors left to establish what became the major tech firm Unisys, the region failed to cultivate a broader computer sector. Arguably, Penn's early leadership should have made the Delaware Valley the first Silicon Valley.[82]

The Sellerses fairly reflected the response of elite Philadelphians to their city and region's descent into economic obscurity. Unlike William, Coleman, and earlier generations, their personal identities were no longer intimately tied to their firms, patents, products, and engineering societies. Like the "Philadelphia gentlemen" studied by sociologist Digby Baltzell, they allied themselves with the achievements of more distant ancestors and with more generic elite social institutions.[83]

Horace's turn to historic preservation augured these changes, as did some of the institutions he frequented. He designed new mansions in the colonial revival style, participating in the paradigm shift from Victorian to historic styles, the architectural equivalent of reversing the changes in machine design made by William.[84] However, when his son Lester wanted to study architecture, Horace insisted that he first get a degree in engineering. Lester joined the family architecture firm just in time for the Great Depression, when the Rockefellers kept them in business overseeing the deconstruction and reconstruction of European monasteries for the Cloisters in New York. Horace, who died in 1933, devoted perhaps his greatest antiquarian energies to his role as official historian of the Sellers Family Association, which he helped found in 1910.[85] He published articles in the

Pennsylvania Magazine of History and Biography on Nathan Sellers, Charles Willson Peale, and other prominent figures of the Revolutionary era.[86]

Later Sellers generations exhibited similar class affiliations. The obituary of William Sellers II focused on his "greatest loves" of sailing, powerboating, hunting, fishing, and flying airplanes, along with his membership in exclusive institutions such as the Merion Cricket Club and Radnor Hunt Club.[87] His second wife, Annah Sellers, was remembered as "a third-generation member of the Radnor Hunt, a former treasurer of the Radnor Horse Trials," vice president of the Philadelphia Society for the Preservation of Landmarks, and president of the Pennsylvania branch of the National Society of the Colonial Dames of America.[88] Social institutions focused on elite Philadelphia's heritage came to define the local bourgeoisie ever more in the twentieth century.

As Baltzell suggested, the Sellerses became part of a "national upper class" of families living off the industrial wealth of cities from Boston to Minneapolis.[89] Their turn-of-the-twentieth-century transformation resulted from what historian John Cumbler termed the replacement of "civic capitalism" characterized by regional networks and institutions in favor of national capitalism.[90] The investment patterns, if perhaps not all the behavior, of Philadelphia's old manufacturing class would be repeated across the country. By and large, the children of late nineteenth-century industrialists pursued service sector professions, often in national corporations, and put their money in the stock market, further centralizing economic power on Wall Street. This generally defined success in the careers and personal wealth of affluent twentieth-century Americans.

In his 1892 tract "American Supremacy in Applied Mechanics," Coleman instructed his readers "that the supremacy is the result of circumstances over which we have some control."[91] Local choices about investment in infrastructure, innovation, and institutions mattered. In the words of historian Philip Scranton, the decline of Philadelphia manufacturing "was not an inexorable corollary to the rise of big business, the ceaseless search for efficiencies, or the intensification of interregional and international competition."[92] It was a messier process, though in the early twentieth century urban and economic changes were increasingly beyond the control of local capitalists. Although not the foremost causes of decline, their decisions and actions helped shape the particular course of decline. In particular, these acts furthered the region's loss of power over key sectors, the erosion of the institutional complex and firm-institution networks, and the transition to a different mode of economic development, one that reflected and sought to respond to a much diminished economic status and competitive position at the national and international levels.

Ironically for manufacturers like the Sellerses, urban reform movements they helped initiate matured to the point where they rationally engineered factories out of the city. Landmark projects like the Fairmount Parkway helped shift Americans' visions and experiences of cities from centers of industrial production to centers of culture, consumption, and service economies. By the mid-twentieth century, Philadelphians earned their metropolis a reputation as a place with an illustrious history but limited economic opportunity, an image and a reality that persist throughout the Rust Belt today.

In the second half of the century, economic development professionals in the public and nonprofit sectors responded with efforts to retrain displaced workers, expand university research, and develop new industrial and office districts. Some of these had important effects, including the survival of some factories that might have closed sooner and the transformation of large parts of the city to new land uses.[93] But none of this appreciably slowed the region's broader deindustrialization, nor did economic development institutions succeed in recapturing Philadelphia's competitive position in the national or world economy.

Every city from Providence and Buffalo to Milwaukee and Kansas City has seen multiple waves of planning and redevelopment aimed at recapturing something of its industrial prosperity. But only Boston, Chicago, and of course New York replaced their old manufacturing economies with robust networks of locally controlled firms and institutions that make them major nodes in the global economy. In recent U.S. Census reports, Philadelphia consistently ranks as one of the poorest big cities in the United States, though around 2008 its population began to grow again.

The boosters and economic development authorities of Philadelphia and other "second tier" cities continually strive to attain the status of global metropolises. With few corporate headquarters locally, many of their strategies center on anchor institutions, mostly immobile medical, cultural, and educational institutions. Their meager success at attracting firms and retaining college graduates highlights the limits within which economic development professionals operate in the hierarchy of world cities. Yet their widespread emphasis on building robust networks of firms and institutions capable of supporting entire sectors acknowledges key drivers of metropolitan growth over the long term. It has also begun to return some measure of innovation and wealth to cities like Philadelphia.

To a great extent, America's old industrial cities retain the physical and institutional landscapes built in the late nineteenth century. Digital innovation notwithstanding, second industrial revolution technologies still structure much of the material environment of cities. Their anchor institutions are largely a product of the institutional complex built by capitalists like the Sellerses. In many ways, so are their strategies of urban planning. Industrial capitalists' multifaceted approach to economic development and social reform remains discernible in urbanists' attempts to engineer prosperity among diverse populations and sectors. The social, natural, and physical sciences they cultivated persist to a great extent as intellectual and material foundations of city building. In these and other ways, industrial history continues to shape the geography and fortunes of the metropolis.

NOTES

—————————————————+—————————————————————————————————————+

Preface

1. In writing his prolific memoirs, Escol corresponded with his brothers about their youth
in the early nineteenth century. See George Escol Sellers to Coleman Sellers, August
14, 1884, and subsequent letters, American Philosophical Society (hereafter APS), Sellers Family Papers. Horace assembled this material plus other letters, family diaries, and
other materials, which his son, prominent twentieth-century historian Charles Coleman
Sellers, gave to the APS.

Introduction

1. Mark Twain and Charles Dudley Warner, *The Gilded Age: A Tale of To-Day* (Hartford,
CT: American Publishing Co., 1873), 83.
2. His full name was George Escol Sellers (Escol without an *h*), though no one called him
George.
3. Charles Dudley Warner to George Escol Sellers, December 23, 1873; J. H. Barton
to George Escol Sellers, December 26, 1873; George Escol Sellers to Charles Dudley
Warner, January 1, 1874; George Escol Sellers to J. H. Barton, March 17, 1875, all
American Philosophical Society (hereafter APS), Sellers Family Papers. For Twain's
apocryphal account of what inspired Warner to name the character Escol, see Charles
Neider, ed., *The Autobiography of Mark Twain* (1917; repr., New York: Perennial, 2000),
26; also, R. Kent Rasmussen, *Mark Twain A to Z* (New York: Facts on File, 1995), 175.
Escol sought for years to shed his association with the fictional colonel. In a meeting
with Warner and the publisher in Hartford later in the winter of 1874, Escol secured
a promise to rename the character. In subsequent editions he became Colonel Beriah
Sellers and later Colonel Mulberry Sellers. But the connection only grew stronger. The
next year, a friend forwarded Escol an "unwarranted personal and libelous article" from
the *Daily Courier* of Evansville, Illinois, alleging he was "the veritable Eschol Sellers"
from the book. As evidence, it noted Sellers' Landing, the town he built to support his

gas and paper ventures, "like Stone's Landing in the book after the failure of the Salt Lick and Pacific Railroad, is deserted, and grass grows in its mud-paved streets." Two weeks later, as the theatrical comedy adapted from the book delighted audiences in Ohio, the *Cincinnati Inquirer* ran the headline, "Colonel Mulberry Sellers—an Old Cincinnatian Identified." The paper labeled him "a real genius of a practical kind, having only two drawbacks, one being that he was far ahead of the age" in his inventions, "and the other that he did not have the proper wealth to carry out his designs to completion; had he had, 'there would have been millions in it,' and to-day the Colonel would have been one of our richest men." As the play's run on Broadway extended into 1878, the *New York Sun* again named him as the prototype for the character in the "alleged novel." G. J. Grammer to George Escol Sellers, March 8, 1875; George Escol Sellers to G. J. Grammer, March 11, 1875; "The Original of Col. Mulberry Sellers: Residing Only Seventy-Five Miles from This City," *Evansville Daily Courier*, March 7, 1875; "Colonel Mulberry Sellers—an Old Cincinnatian Identified," *Cincinnati Inquirer*, March 26, 1875; and George Escol Sellers to J. H. Barton, September 8, 1878; George Escol Sellers to editor of the *New York Sun*, September 9, 1878; J. H. Barton to George Escol Sellers, September 13, 1878; Coleman Sellers to George Escol Sellers, September 18, 1878; J. H. Barton to George Escol Sellers, September 22, 1878, all APS, Sellers Family Papers. "Twain and Sellers: How Clemens and Warner Came to Write Their Alleged Novel," *New York Sun*, August 6, 1878; John W. Maxson Jr., "George Escol Sellers: Inventor, Historian, and Papermaker," *Papermaker* 38, no. 1 (1969): 40–57. Into the 1890s, newspapers continued to run profiles that linked him to *The Gilded Age*; see, for example, "A True Story of Col. Mulberry Sellers Now Living in Chattanooga," *Chattanooga Republican*, March 8, 1891, and Escol's response, "Mark Twain Says That Eschol Sellers Was Not the Original 'Mulberry,'" *Chattanooga Republican*, March 15, 1891.

4. Escol's reminiscences appeared periodically in *American Machinist*, vols. 7–18 (1884–95); selections appear in Eugene Ferguson, ed., *Early Engineering Reminiscences (1815–1840) of George Escol Sellers* (Washington, DC: Smithsonian Institution, 1965). See also Horace Wells Sellers, in *Report of the Proceedings of the Third Meeting of the Sellers Family Association* (Philadelphia: Leeds & Biddle, 1913), 8.

5. For seminal works that have defined the scope of pre-twentieth-century planning history in relatively narrow terms, not explicitly including economic development (despite noting the economic interests of early planners and their employers), see John W. Reps, *The Making of Urban America: A History of City Planning in the United States* (Princeton, NJ: Princeton University Press, 1992); Jon Peterson, *The Birth of City Planning in the United States, 1840–1917* (Baltimore: Johns Hopkins University Press, 2003).

6. The broadening scope of planning historiography in recent decades is examined in Robin Bachin, "City Building as Community Building: Re-visioning Planning History," *Journal of Planning History* 1, no. 3: 235–39; and Domenic Vitiello, "Re-forming Schools and Cities: Placing Education on the Landscape of Planning History," *Journal of Planning History* 5, no. 3 (2006): 183–95. Recent works in planning history that have expanded upon various dimensions of economic development in North America, especially industrial districts' role in shaping metropolitan growth, include Greg Hise, *Magnetic Los Angeles: Planning the Twentieth-Century Metropolis* (Baltimore: Johns Hopkins University Press, 1999); Robert Lewis, ed., *Manufacturing Suburbs: Building Work and Home on the Metropolitan Fringe* (Philadelphia: Temple University Press, 2004); Robert Lewis, *Manufacturing Montreal: The Making of an Industrial Landscape, 1850 to 1930* (Baltimore: Johns Hopkins University Press, 2000); Robert Lewis, *Chicago Made: Factory Networks in the Industrial Metropolis* (Chicago: University of Chicago Press, 2008). See also the encyclopedic Paul Bairoch, *Cities and Economic Development: From the Dawn of History to the Present*, trans. Christopher Braider (Chicago: University of Chicago Press, 1991); and Thomas Dublin and Walter Licht, *The Face of Decline: The Pennsylvania Anthracite Region in the Twentieth Century* (Ithaca, NY: Cornell University Press, 2005). For urban and planning history literature that may be read implicitly as history of economic

development, see for example Reps, *Making of Urban America*; Stephen V. Ward, *Planning the Twentieth-Century City: The Advanced Capitalist World* (New York: Wiley, 2002); Kenneth Jackson, "The Capital of Capitalism: The New York Metropolitan Region, 1890–1940," in *Metropolis, 1890–1940*, ed. Anthony Sutcliffe (Chicago: University of Chicago Press, 1984), 319–54; and also other chapters in this volume.

7. Intermediaries were critical to nineteenth-century social theorists' conceptions of industrialization, and they occupy a similarly critical place in the models of many contemporary historians and sociologists. The roles of engineers in shaping the Enlightenment and early industrial society are treated in Ken Alder, *Engineering the Revolution: Arms and Enlightenment in France, 1763–1815* (Princeton, NJ: Princeton University Press, 1997). On the social roles of engineers in later periods, see Monte Calvert, *The Mechanical Engineer in America, 1830–1910* (Baltimore: Johns Hopkins University Press, 1967); Daniel H. Calhoun, *The American Civil Engineer: Origins and Conflict* (Cambridge, MA: MIT Press, 1960); David F. Noble, *America by Design: Science, Technology, and the Rise of Corporate Capitalism* (New York: Alfred A. Knopf, 1977); Thomas Hughes, *Networks of Power: Electrification in Western Society, 1880–1930* (Baltimore: Johns Hopkins University Press, 1983); Thomas J. Misa, *A Nation of Steel: The Making of Modern America, 1865–1925* (Baltimore: Johns Hopkins University Press, 1995); Ross Thomson, ed., *Learning and Technological Change* (New York: St. Martin's, 1993); Wiebe Bijker, Thomas P. Hughes, and Trevor J. Pinch, eds., *The Social Construction of Technological Systems: New Directions in the Sociology and History of Technology* (Cambridge, MA: MIT Press, 1987); Wiebe Bijker and John Law, eds., *Shaping Technology / Building Society: Studies in Sociotechnical Change* (Cambridge, MA: MIT Press, 1992).

8. For recent overviews and assessments of this literature, see for example Guian McKee, *The Problem of Jobs: Liberalism, Race, and Deindustrialization in Philadelphia* (Chicago: University of Chicago Press, 2008); Howard Gillette, *Camden after the Fall: Decline and Renewal in a Postindustrial City* (Philadelphia: University of Pennsylvania Press, 2005); Jefferson Cowie and Joseph Heathcott, eds., *Beyond the Ruins: The Meanings of Deindustrialization* (Ithaca, NY: Cornell University Press, 2003); Thomas J. Sugrue, *The Origins of the Urban Crisis: Race and Inequality in Postwar Detroit* (Princeton, NJ: Princeton University Press, 1996). Studies that stress the early twentieth-century structural roots of post–World War II deindustrialization include John Cumbler, *A Social History of Economic Decline: Business, Politics, and Work in Trenton* (New Brunswick, NJ: Rutgers University Press, 1989); Philip Scranton, "Large Firms and Industrial Restructuring: The Philadelphia Region 1900–1980," *Pennsylvania Magazine of History and Biography* 116 (1992): 419–65. I have elaborated this argument in Domenic Vitiello, "Machine Building and City Building: The Planning and Politics of Urban and Industrial Restructuring in Philadelphia, 1891–1928," *Journal of Urban History* 34, no. 3 (March 2008): 399–434.

9. Historical and social science literature that emphasizes the importance of networks includes David Meyer, *The Roots of American Industrialization* (Baltimore: Johns Hopkins University Press, 2003), and *Hong Kong as a Global Metropolis* (New York: Cambridge University Press, 2000), esp. 6–7; Manuel Castells, *The Rise of the Network Society* (Oxford: Blackwell, 2000); Harrison C. White, *Markets from Networks: Socioeconomic Models of Production* (Princeton, NJ: Princeton University Press, 2002); Eric Sheppard and Trevor Barnes, *A Companion to Economic Geography* (Oxford: Blackwell, 2000); Ulrich Jurgens, ed., *New Product Development and Production Networks: Global Industrial Experience* (New York: Springer, 2000); Immanuel Wallerstein and Terence K. Hopkins, *Processes of the World-System* (Beverly Hills, CA: Sage, 1980). See also Linton Freeman, *The Development of Social Network Analysis* (Vancouver: Empirical Press, 2006).

10. Lewis, *Chicago Made* and *Manufacturing Montreal*; Lewis, *Manufacturing Suburbs*, including Edward Muller, "Industrial Suburbs and the Growth of Metropolitan Pittsburgh, 1870–1920," 124–42, and Greg Hise, " 'Nature's Workshop': Industry and Urban Expansion in Southern California, 1900–1950," 178–99; and Hise, *Magnetic Los Angeles*.

11. Muller, "Industrial Suburbs," 125.

12. Lewis, *Chicago Made*, 14.

13. Works that detail metropolitan industrial networks' external dynamics include William Cronon, *Nature's Metropolis: Chicago and the Great West* (New York: W. W. Norton, 1991); Lewis, *Chicago Made*; Hise, *Magnetic Los Angeles*; Meyer, *Roots of American Industrialization*.

14. The classic study of economic geography in industrial America is Allan R. Pred, *City Systems in Advanced Economies: Past Growth, Present Processes, and Future Development Options* (New York: Wiley, 1977); see also S. N. Eisenstadt and A. Sachar, *Society, Culture, and Urbanization* (Beverly Hills, CA: Sage, 1987); Cronon, *Nature's Metropolis*; David Meyer, "A Dynamic Model of the Integration of Frontier Urban Places into the United States System of Cities," *Economic Geography* 56 (April 1980): 120–40, "Emergence of the American Manufacturing Belt: An Interpretation," *Journal of Historical Geography* 9, no. 2 (1983): 145–74, "The Rise of the Industrial Metropolis: The Myth and the Reality," *Social Forces* 68 (March 1990): 731–52, "The National Integration of Regional Economies, 1860–1920," in *North America: The Historical Geography of a Changing Continent*, ed. Thomas F. McIlwraith and Edward K. Muller (Lanham, MD: Rowman & Littlefield, 2001), 307–31, and *Roots of American Industrialization*; Gordon M. Winder, "The North American Manufacturing Belt in 1880: A Cluster of Regional Industrial Systems or One Large Industrial District?" *Economic Geography* 75, no. 1 (1999): 71–92; Nathan Rosenberg and L. E. Birdzell Jr., *How the West Grew Rich: The Economic Transformation of the Industrial World* (New York: Basic Books, 1986); Lewis, *Manufacturing Suburbs*. For case and regional studies of industrial geography, see Ann R. Markusen, Peter Hall, Scott Campbell, and Sabrina District, *The Rise of the Gunbelt: The Military Remapping of Industrial America* (New York: Oxford University Press, 1991); Roger Lotchin, *Fortress California, 1910–1961: From Warfare to Welfare* (New York: Oxford University Press, 1992); Hise, *Magnetic Los Angeles*; Lewis, *Chicago Made* and *Manufacturing Montreal*.

15. For overviews of the "new institutional economics," see Richard N. Langlois, ed., *Economics as a Process: Essays in the New Institutional Economics* (New York: Cambridge University Press, 1986); Douglass C. North, *Institutions, Institutional Change, and Economic Performance* (New York: Cambridge University Press, 1990); John N. Drobak and John V. C. Nye, eds., *The Frontiers of the New Institutional Economics* (San Diego: Academic, 1997).

16. For example, Michael B. Katz, *Improving Poor People: The Welfare State, the "Underclass," and Urban Schools as History* (Princeton, NJ: Princeton University Press, 1997); Paul Boyer, *Urban Masses and Moral Order in America, 1820–1920* (Cambridge, MA: Harvard University Press, 1978); Kim Phillips-Fine, *Invisible Hands: The Businessmen's Crusade against the New Deal* (New York: W. W. Norton, 2010).

17. For example, Lawrence Vale, *From the Puritans to the Projects: Public Housing and Public Neighbors* (Cambridge: Harvard University Press, 2000); John F. Bauman, *Public Housing, Race, and Renewal: Urban Planning in Philadelphia, 1920–1974* (Philadelphia: Temple University Press, 1974); David Schuyler, *The New Urban Landscape: The Redefinition of City Form in Nineteenth-Century America* (Baltimore: Johns Hopkins University Press, 1986); John Teaford, *The Rough Road to Renaissance: Urban Revitalization in America, 1940–1985* (Baltimore: Johns Hopkins University Press, 1990); McKee, *Problem of Jobs*; Hise, " 'Nature's Workshop' "; Lewis, *Chicago Made*; Cronon, *Nature's Metropolis*; Richard Walker, *The Country in the City: The Greening of the San Francisco Bay Area* (Seattle: University of Washington Press, 2007). Works that break down the dichotomy between public and private institutions in social, legal, and industrial and environmental history include Michael B. Katz, *The Price of Citizenship* (New York: Metropolitan, 2001); William Novak, *The People's Welfare: Law and Regulation in Nineteenth-Century America* (Chapel Hill: University of North Carolina Press, 1996); Theodore Steinberg, *Nature Incorporated: Industrialization and the Waters of New England* (New York: Cambridge University

Press, 1991). Some historians have studied metropolitan complexes of institutions, even if they do not frame their analysis in these terms. One of the most focused analyses of institutions' collective impacts on urban life is Thomas Bender, *Toward an Urban Vision: Ideas and Institutions in Nineteenth-Century America* (Baltimore: Johns Hopkins University Press, 1992).

18. The impacts of institutional "interactivity" on industrialization are discussed in Philip Scranton, *Endless Novelty: Specialty Production and American Industrialization, 1865–1925* (Princeton, NJ: Princeton University Press, 1997), 157–60. The effects of "interactivity" upon urbanization are discussed in Jane Jacobs, *The Economy of Cities* (New York: Random House, 1969), 85–89; Jane Jacobs, *Cities and the Wealth of Nations: Principles of Economic Life* (New York: Random House, 1984), 40–41.

19. For histories of engineering, architecture, urban planning, and the environment focused on engineers' roles in transforming the ecology of industrial society, see for example Tom F. Peters, *Building the Nineteenth Century* (Cambridge, MA: MIT Press, 1996); George E. Thomas, *William L. Price: From Arts and Crafts to Modern Design* (New York: Princeton Architectural Press, 2000); Clifton Hood, *772 Miles: The Building of the Subways and How They Transformed New York* (Baltimore: Johns Hopkins University Press, 1993); Richard G. Wilkinson, "The English Industrial Revolution," in *The Ends of the Earth*, ed. Donald Worster (New York: Cambridge University Press, 1988), 80–102; E. A. Wrigley, *Continuity, Chance and Change: The Character of the Industrial Revolution in England* (New York: Cambridge University Press, 1988); Cronon, *Nature's Metropolis*; Richard White, *The Organic Machine: The Remaking of the Columbia River* (New York: Hill & Wang, 1995); Adam Rome, *The Bulldozer in the Countryside: Suburban Sprawl and the Rise of American Environmentalism* (New York: Cambridge University Press, 2001).

20. Recent literature that traces the ways in which people's decisions influenced the course of deindustrialization in particular places includes Jefferson Cowie, *Capital Moves: RCA's Seventy-Year Quest for Cheap Labor* (Ithaca, NY: Cornell University Press, 1999); Cumbler, *A Social History of Economic Decline*; Scranton, "Large Firms and Industrial Restructuring"; Cowie and Heathcott, *Beyond the Ruins*.

1. Manufacturing Metropolitan Development

1. Nathan & David Sellers, advertisement (n.d.), Sellers Memorial Library.
2. Nathan & David Sellers, Account Book, 1774–1801, APS, Peale-Sellers Papers.
3. Certificates of his membership are found in the Peale-Sellers Papers, APS.
4. Nathan Sellers, Notes from Common Council, 1806–12, APS, Peale-Sellers Papers.
5. George Escol Sellers, "Personal Recollections of Nathan Sellers by His Grandson George Escol Sellers," January 12, 1894, APS, Sellers Family Papers.
6. The context of migration from the North Midlands is described in David Hackett Fischer, *Albion's Seed: Four British Folkways in America* (New York: Oxford University Press, 1989), 445–51.
7. Ellen Cronin, "The Sellers Family of Upper Darby: An Eighteenth and Nineteenth Century American Dynasty of Inventors, Millers and Mechanic-Manufacturers" (typescript in the possession of the author).
8. William Penn, "A Further Account of the Province of Pennsylvania, 1685," in *Narratives of Early Pennsylvania, West New Jersey, and Delaware, 1630–1707*, ed. Albert Cook Meyers (New York: Charles Scribner's Sons, 1912), 263. See also Sylvia Doughty Fries, *The Urban Idea in Colonial America* (Philadelphia: Temple University Press, 1977).
9. Nicholas Sellers, Peter H. Sellers, Lucy Bell Sellers, William W. Sellers, and Nancy Sellers, "Sellers Tricentennial, 1681–1981," 1–2, Historical Society of Pennsylvania (hereafter HSP), Philadelphia. On the cultural, social, and ecological significance of clearing land, marking boundaries, building houses, and building European-style agriculture

and settlements, see Patricia Seed, *Ceremonies of Possession in Europe's Conquest of the New World, 1492–1640* (New York: Cambridge University Press, 1995); William Cronon, *Changes in the Land: Indians, Colonists, and the Ecology of New England* (New York: Hill & Wang, 1983); Alfred W. Crosby, *Ecological Imperialism: The Biological Expansion of Europe, 900–1900* (New York: Cambridge University Press, 1986).

10. Darby Monthly Meeting Minutes, May 2, 1684, Friends Historical Library, Swarthmore College (hereafter FHL).

11. Nicholas Sellers et al., "Sellers Tricentennial," 4–5; Henry Graham Ashmead, *History of Delaware County, Pennsylvania* (Philadelphia: Everts, 1884), 532; George Smith, *History of Delaware County, Pennsylvania* (Philadelphia: Ashmead, 1862), 499.

12. Stevenson Whitcomb Fletcher, *Pennsylvania Agriculture and Country Life, 1640–1840* (Harrisburg, PA: PHMC, 1971), 324.

13. Michal McMahon, "'Publick Service' versus 'Mans Properties': Dock Creek and the Origins of Urban Technology in Eighteenth-Century Philadelphia," in *Early American Technology: Making and Doing Things from the Colonial Era to 1850*, ed. Judith A. McGaw (Chapel Hill: University of North Carolina Press, 1994), 114–47.

14. Smith, *History of Delaware County*, 499; Whitfield J. Bell Jr., *Patriot-Improvers: Biographical Sketches of Members of the American Philosophical Society*, vol. 1, *1743–1768* (Philadelphia: APS, 1997), 500.

15. John W. Maxson Jr., "Nathan Sellers: America's First Large-Scale Maker of Paper Moulds," *Paper Maker* 29, no. 1 (1960): 1–15; Nicolas Sellers et al., "Sellers Tricentennial," 4–5.

16. *Millbourne Mills Company: Its Antecedents and Present Operations* (Philadelphia, 1888), 6. Also, S. F. Hotchkin, *Rural Pennsylvania: In the Vicinity of Philadelphia* (Philadelphia: Jacobs, 1897), 309; Ashmead, *History of Delaware County*, 547.

17. *Pennsylvania Gazette*, September 3, 1767; Bell, *Patriot-Improvers*, 1:501. Correspondence survives between John Sellers and customers as far away as Albany, New York. John Sellers to George Reale, July 17, 1789, APS, Peale-Sellers Papers.

18. Nicholas Sellers et al., "Sellers Tricentennial," 6.

19. Nathan Sellers, "Common Place Book," beginning May 29, 1772, APS, Sellers Family Papers.

20. Nicholas Sellers et al., "Sellers Tricentennial," 9.

21. Ken Alder calls this a "technological life," "a coherent social and ideological world which gives purpose and meaning to a set of material objects." Alder, *Engineering the Revolution: Arms and Enlightenment in France, 1763–1815* (Princeton, NJ: Princeton University Press, 1997), xii. Also, Laura Rigal, *The American Manufactory: Art, Labor, and the World of Things in the Early Republic* (Princeton, NJ: Princeton University Press, 1998).

22. Nathan Sellers, "Common Place Book"; and Nathan Sellers, Diary, 1806–1817), APS, Peale-Sellers Papers.

23. Nathan Sellers, Diary, November 1817.

24. John Sellers, Survey of Carpenters Island (May 1764), HSP, Gratz Collection, Miscellaneous Business Papers, case 15, box 21; John Sellers et al. to Richard Penn, June 5, 1773, HSP, Penn Manuscripts, Warrants & Surveys, large folio, 75; Bell, *Patriot-Improvers*, 1:502.

25. Cronon, *Changes in the Land*.

26. Histories of the Philosophical Society include Murphy D. Smith, *Oak from an Acorn: A History of the American Philosophical Society Library, 1770–1804* (Wilmington, DE: Scholarly Resources, 1976); and Peter Du Ponceau, *An Historical Account of the Origin and Formation of the American Philosophical Society Held at Philadelphia for Promoting Useful Knowledge* (Philadelphia: APS, 1914). For the University of Pennsylvania, see George E. Thomas and David B. Brownlee, *Building America's First University: An Historical and Architectural Guide to the University of Pennsylvania* (Philadelphia: University of

Pennsylvania Press, 2000). On John's contribution to the hospital, see Bell, *Patriot-Improvers*, 1:502.

27. *Early Proceedings of the American Philosophical Society for the Promotion of Useful Knowledge . . . 1744 to 1838* (Philadelphia: McCalla & Stavely, 1884), 1–6, APS.

28. Whitfield J. Bell Jr., *Patriot-Improvers: Biographical Sketches of Members of the American Philosophical Society*, vol. 2, *1768* (Philadelphia: APS, 1999), 220–21; Anna T. Lincoln, *Wilmington, Delaware, Three Centuries under Four Flags, 1609–1937* (Port Washington, NY: Kennikat, 1972), 106, 217.

29. *Early Proceedings of the American Philosophical Society*, 30.

30. Ibid., 41.

31. Ibid., 36; Bell, *Patriot-Improvers*, 2:220.

32. Hotchkin, *Rural Pennsylvania*, 328; Nicholas Sellers et al., "Sellers Tricentennial," 8; Bell, *Patriot-Improvers*, 1:502.

33. Nathan Sellers, Diary, 1776.

34. George Escol Sellers, Memoirs, vol. 1, p. 1, APS, Peale-Sellers Papers; Samuel Breck, "Historical Sketch of the Continental Bills of Credit, from the Year 1775 to 1781," 1856, APS, Peale-Sellers Papers.

35. David McCullough, *John Adams* (New York: Simon & Schuster, 2001), 141.

36. Quoted in Maxson, "Nathan Sellers," 5.

37. Ibid., 6; Nathan & David Sellers, Account Book; M. Hillegas to Nathan Sellers, April 11, 1778, APS, Sellers Family Papers.

38. Quoted in Smith, *History of Delaware County*, 299.

39. Bell, *Patriot-Improvers*, 1:502.

40. Nathan Sellers, Diary, March 22, 1818.

41. Nathan Sellers, Surveying Data (1777–84), APS, Sellers Family Papers.

42. David Rittenhouse, Thomas Hutchins, and Nathan Sellers, *Essay or Report to the Genl. Assembly of Pennsylvania on the Union of the Waters of Susquehanna & Schuylkill by the Tulpehocken & Swatara Creeks* (1784), APS, Sellers Family Papers.

43. *Proceedings of the Fourth and Fifth General Meetings of the Sellers Family Association Held in the Years 1916 and 1920* (Philadelphia: Leeds & Biddle, 1920), 11.

44. Bell, *Patriot-Improvers*, 1:503.

45. Address of the Senate to Gov. Thomas Mifflin, August 26, 1791, Library Company of Philadelphia (hereafter LCP).

46. Journal of the Society for the Improvement of Roads and Inland Navigation (1791–93), HSP.

47. On the Federalist elite's attempt to retain control of the city's economy through financial institutions and state-chartered corporations, see Andrew M. Schocket, "Consolidating Power: Technology, Ideology, and Philadelphia's Growth in the Early Republic" (dissertation, College of William and Mary, 2001).

48. On Federal-era Philadelphia's financial sector and its impacts on regional economic development, see Domenic Vitiello with George E. Thomas, *The Philadelphia Stock Exchange and the City It Made* (Philadelphia: University of Pennsylvania Press, 2010), chap. 2.

49. Diary of John Sellers Jr., November 19, 1832, and November 18, 1833, APS, Peale-Sellers Papers.

50. Nathan Sellers, notebook detailing investments, 1815–1844, APS, Sellers Family Papers.

51. Bell, *Patriot-Improvers*, 1:503; John Sellers, "Answers to Queries on Plaister of Paris," in Richard Peters, *Agricultural Inquiries on Plaister of Paris* (Philadelphia, 1810), 46–52.

52. Simon Baatz, *"Venerate the Plough": A History of the Philadelphia Society for Promoting Agriculture, 1785–1985* (Philadelphia: PSPA, 1985), 21–26.

53. David R. Brigham, *Public Culture in the Early Republic: Peale's Museum and Its Audience* (Washington, DC: Smithsonian Institution Press, 1995), 1.

54. Charles Coleman Sellers, *Mr. Peale's Museum: Charles Willson Peale and the First Popular Museum of Natural Science and Art* (New York: W. W. Norton, 1980), 1, 82, 239. See also Edgar P. Richardson et al., *Charles Willson Peale and His World* (New York: Abrams, 1982).

55. George Escol Sellers to Horace Wells Sellers, May 2, 1895, Peale-Sellers Papers, APS; Sellers, *Mr. Peale's Museum*, 242–44.

56. Sellers, "Personal Recollections of Nathan Sellers."

57. William Novak, *The People's Welfare: Law and Regulation in Nineteenth-Century America* (Chapel Hill: University of North Carolina Press, 1996); Oscar Handlin and Mary Handlin, *Commonwealth: A Study of the Role of Government in the American Economy: Massachusetts, 1774–1861* (New York: NYU Press, 1947).

58. Sellers, Notes from Common Council.

59. Nathan Sellers, notebook detailing investments (1815–1844). Year 2010 dollars were calculated according to the Consumer Price Index (for this and all subsequent dollar value comparisons), on the Measuring Worth website: http://www.measuringworth. com/uscompare/ (visited June 27, 2011).

60. Sellers, Notes from Common Council.

61. Philip Scranton, *Proprietary Capitalism: The Textile Manufacture at Philadelphia, 1800–1885* (New York: Cambridge University Press, 1983), 75. Most of this increase was the result of migration from the city's rural hinterlands, as metropolitan economic growth was largely driven by demand for raw materials and manufactured goods from a relatively prosperous farming population as well as urban households, manufacturers, and transportation companies within the region (overseas migration and trade accounted for a small proportion of this overall growth). This intra-regional demand model for Philadelphia is explained in Diane Lindstrom, *Economic Development in the Philadelphia Region, 1810–1850* (New York: Columbia University Press, 1978), and for the North and East in David Meyer, *The Roots of American Industrialization* (Baltimore: Johns Hopkins University Press, 2003). See also Bruce Laurie, *Working People of Philadelphia, 1800–1850* (Philadelphia: Temple University Press, 1980), 9–10; Burton W. Folsom Jr., *Urban Capitalists: Entrepreneurs and City Growth in Pennsylvania's Lackawanna and Lehigh Regions, 1800–1920*, 2nd ed. (Scranton, PA: University of Scranton Press, 2000); Donna J. Rilling, *Making Houses, Crafting Capitalism: Builders in Philadelphia, 1790–1850* (Philadelphia: University of Pennsylvania Press, 2001).

62. Sellers, Notes from Common Council. On the waterworks' engineering, see also Martin Melosi, *The Sanitary City: Urban Infrastructure in America from Colonial Times to the Present* (Baltimore: Johns Hopkins University Press, 2000), 31–34. On early government attempts to regulate industrial pollution in Philadelphia, see McMahon, "'Publick Service' versus 'Mans Properties.'"

63. Sellers, "Personal Recollections of Nathan Sellers."

64. Certificates of Nathan's membership in these institutions are in the Peale-Sellers Papers, APS; Richard Newman, "The Pennsylvania Abolition Society: Restoring a Group to Glory," *Pennsylvania Legacies*, November 2005, 8.

65. Nathan Sellers to Elizabeth Coleman, winter 1778–79, Peale-Sellers Papers, APS.

66. Newman, "Pennsylvania Abolition Society," 8.

67. Sellers, "Personal Recollections of Nathan Sellers."

68. Nathan Sellers to Lehman, January 23, 1822, APS, Sellers Family Papers. See also Nathan Sellers to Mathew Carey, Nicholas Biddle, et al., October 23, 1824, LCP, M. Carey, Correspondence on Internal Improvements; Mathew Cary to Nathan Sellers, December 6, 1824, APS, Sellers Family Papers; William Strickland, *Reports on Canals, Railways and Other Subjects* (Philadelphia, 1826).

69. On the failures of Pennsylvania's internal improvements, see John Lauritz Larson, *Internal Improvement: National Public Works and the Promise of Popular Government in the Early United States* (Chapel Hill: University of North Carolina Press, 2001). For a counterargument, see Meyer, *Roots of American Industrialization*, esp. chaps. 2 and 5; also, Donald

C. Jackson, "Roads Most Traveled: Turnpikes in Southeastern Pennsylvania in the Early Republic," in McGaw, *Early American Technology*, 197–239. For analysis of Philadelphia stockbrokers' investments in interurban infrastructure and in other cities, see Vitiello, *Philadelphia Stock Exchange*, chaps. 3 and 4.

70. For a review of historians' interpretations of industrialization, see Pat Hudson, *The Industrial Revolution* (New York: Oxford University Press, 1992).

71. Nathan & David Sellers, advertisement.

72. Anthony F. C. Wallace, *Rockdale: The Growth of an American Village in the Early Industrial Revolution* (New York: Alfred A. Knopf, 1978), 211.

73. Nathan & David Sellers, Account Book.

74. Nathan & David Sellers, Account Book; Maxson, "Nathan Sellers," 9–12; Charles Bird, Bills and Orders, April 9, 1818, and two orders undated, LCP.

75. Lawrence A. Peskin, *Manufacturing Revolution: The Intellectual Origins of Early American Industry* (Baltimore: Johns Hopkins University Press, 2003), 103.

76. Sellers, Memoirs, vol. 1, p. 43.

77. Ashmead, *History of Delaware County*, 546.

78. Nicholas Sellers et al., "Sellers Tricentennial," 10.

79. *Millbourne Mills Company*, 7.

80. Samuel Sellers to John Sellers, February 20, 1811, APS, Sellers Family Papers.

81. John Sellers to Nathan Sellers, 1820, APS, Sellers Family Papers.

82. Ashmead, *History of Delaware County*, 547.

83. *Early Proceedings of the American Philosophical Society*, 428.

84. John W. Maxson Jr., "Coleman Sellers: Machine Maker to America's First Mechanized Paper Mills," *Paper Maker* 30, no. 1 (1961): 13–27.

85. Nathan Sellers to John Quincy Adams, July 21, 1820; see also Nathan Sellers to William Thornton, July 21, 1820; Nathan Sellers, Affirmation of Originality for Patent, (July 21, 1820); Nathan Sellers, "Specification of an Improvement in Cleaning Sifting Seperating Dressing and Polishing Shelled Pounded or Scoured Rice and of Cleaning or Sanding Rough Rice" (July 21, 1820); Nathan Sellers, "Synthetical View of the Laws of Patents" (July 1820); William Thornton to Nathan Sellers, July 29, 1820; Nathan Sellers to John Quincy Adams, July 29, 1820; Nathan Sellers to William Thornton, August 1, 1820, all APS, Sellers Family Papers.

86. Sellers, Memoirs, vol. 1, p. 47.

87. Nathan Sellers, Diary, June 1817.

88. Sellers, "Personal Recollections of Nathan Sellers."

89. Joseph Jackson, *Market Street: The Most Historic Highway in America* (Philadelphia: Jackson, 1918), 98.

90. Wallace, *Rockdale*, 216.

91. Pennock and David's son James Sellers patented the process of riveting leather, though it was developed with the aid of Coleman and his father-in-law, Charles Willson Peale.

92. George Escol Sellers, "Recollections of Coleman Sellers by his Son George Escol Sellers," 48–49, APS, Peale-Sellers Papers. The emphasis is in original.

93. Wallace, *Rockdale*, 213.

94. Eugene Ferguson, *Oliver Evans: Inventive Genius of the American Industrial Revolution* (Wilmington, DE: Hagley Museum and Library, 1980).

95. Eugene Ferguson, ed., *Early Engineering Reminiscences (1815–1840) of George Escol Sellers* (Washington, DC: Smithsonian Institution, 1965), 20.

96. For energy-centered interpretations of the industrial revolution, see Richard G. Wilkinson, "The English Industrial Revolution," in *The Ends of the Earth: Perspectives on Modern Environmental History*, ed. Donald Worster (New York: Cambridge University Press, 1988), 80–99; and E. A. Wrigley, *Continuity, Chance and Change: The Character of the Industrial Revolution in England* (New York: Cambridge University Press, 1988).

97. Josiah White, *Josiah White's History Given by Himself* (Philadelphia: Lehigh Coal & Navigation Co., 1909), 32.

98. Ibid. See also Norris Hansell, *Josiah White: Quaker Entrepreneur* (Easton, PA: Canal History and Technology Press, 1992).

99. Nathan Sellers, notebook detailing investments (1815–1844).

100. Ferguson, *Early Engineering Reminiscences*, 4.

101. Hotchkin, *Rural Pennsylvania*, 332–34; Ferguson, *Early Engineering Reminiscences*, 6, 43–44; Maxson, "Coleman Sellers"; James Sellers & Abraham Pennock, Letters Patent, July 6, 1818, HSP; Coleman Sellers to Samuel Sellers, January 15, 1822, and John J. Peabody to William Sellers & Co., August 16, 1897, both APS, Peale-Sellers Papers.

102. Ferguson, *Early Engineering Reminiscences*, 43–44, 49; see also David Freeman Hawke, *Nuts and Bolts of the Past: A History of American Technology, 1776–1860* (New York: Harper & Row, 1988), 37.

103. Charles Sellers to George Escol Sellers, April 18, 1884, APS, Sellers Family Papers.

104. Sellers, Brandt & Co., Letter Book (1828–1834), APS, Peale-Sellers Papers.

105. Quoted in John W. Maxson Jr., "George Escol Sellers: Inventor, Historian, and Papermaker," *Paper Maker* 38, no. 1 (1969): 42.

106. Wallace, *Rockdale*, 147–48.

107. W. Harrold to Charles and George Escol Sellers, October 21, 1834, APS, Peale-Sellers Papers.

108. Sellers, Memoirs, vol. 23, p. 38.

109. Ibid., vol. 1, p. 16.

110. Wallace, *Rockdale*, 210.

111. By midcentury some 43 percent of Philadelphia's industrial workers toiled in factories with more than fifty employees. Laurie, *Working People of Philadelphia*, 14.

112. Meyer, *Roots of American Industrialization*, 126.

113. Ashmead, *History of Delaware County*, 546.

114. Wallace, *Rockdale*, 42, 89–91.

115. Nathan Sellers, notebook detailing investments (1815–1844).

116. Wallace, *Rockdale*, 17, also 24, 100.

117. Coleman Sellers & Sons, Order Book (1834–36), and Robert M. Patterson to Charles & Escol Sellers, October 6, 1836, both APS, Peale-Sellers Papers; Coleman and George Escol Sellers business/legal papers (1834–45), HSP, Cadwalader Collection, Series 6, box 103, folder 4; Ferguson, *Early Engineering Reminiscences*, 62; Hotchkin, *Rural Pennsylvania*, 332; John H. White, *Cincinnati Locomotive Builders, 1845–1868* (Washington, DC: Smithsonian Institution Press, 1965), 50.

118. *Henry Troth, Sept. 4, 1794–May 22, 1842* (Philadelphia, 1903); Pennsylvania Society for the Promotion of Public Economy, *Report of the Library Committee of the Pennsylvania Society for the Promotion of Public Economy* (Philadelphia, 1817), 39.

119. Michael Meranze, *Laboratories of Virtue: Punishment, Revolution, and Authority in Philadelphia, 1760–1835* (Chapel Hill: University of North Carolina Press, 1996), 237.

120. *Laws of the Commonwealth of Pennsylvania*, vol. 7 (Philadelphia: John Bioren, 1822), 53–59; quote is from 58.

121. Michael B. Katz, *Reconstructing American Education* (Cambridge, MA: Harvard University Press, 1987), 13. The emphasis is in original.

122. Peskin, *Manufacturing Revolution*, esp. 155.

123. Job R. Tyson, *An Address Delivered at the Request of the Board of Managers of the Apprentices' Library Company of Philadelphia* (Philadelphia: Young, 1830).

124. Thomas and Brownlee, *Building America's First University*, 48.

125. The quote is from *Constitution and By-Laws of the Franklin Institute* (Philadelphia: Barnard & Jones, 1854), 6. Wallace, *Rockdale*, 229; Election of Board of Managers, February 16, 1824, Franklin Institute, Minutes of the Board of Managers.

126. *Constitution and By-Laws of the Franklin Institute*, 24. See also Percy A. Bivins, *Index to the Reports of the Committee on Science and the Arts* (Philadelphia: Franklin Institute, 1890);

A. Michal McMahon and Stephanie A. Morris, *Technology in Industrial America: The Committee on Science and the Arts of the Franklin Institute, 1824–1900* (Wilmington, DE: Scholarly Resources, 1977).

127. P. Thomas Carroll, "The Decline of the American Philosophical Society, 1820–1855: A Prosopographical First Look at Some Changes in American Science during the First Half of the Nineteenth Century" (University of Pennsylvania, Department of History and Sociology of Science, paper, February 1974), 14, 18, APS. See also J. Peter Lesley, "The Spirit of a Philosophical Society," *Proceedings of the American Philosophical Society* 18 (1878–80): 586–87; Whitfield J. Bell Jr., "As Others Saw Us: Notes on the Reputation of the American Philosophical Society," *Proceedings of the American Philosophical Society* 116 (1972): 269–78; Walter Elliot Gross, *The American Philosophical Society and the Growth of Science in the United States, 1835–1850* (dissertation, University of Pennsylvania, 1970).

128. Samuel V. Merrick, *Report upon an Examination of Some of the Gas Manufactories in Great Britain, France, and Belgium, under a Resolution Passed by the Select and Common Councils of the City of Philadelphia* (Pittsburgh: Hogan, 1835); *Report of the Managers of the Franklin Institute . . . in Relation to Weights and Measures* (Philadelphia: Harding, 1834); Bruce Sinclair, *Early Research at the Franklin Institute: The Investigation into the Causes of Steam Boiler Explosions, 1830–1837* (Philadelphia: Franklin Institute, 1966).

129. Coleman Sellers, "Mechanics, an Introductory Lecture Delivered before the Franklin Institute" (Philadelphia, 1884), 3; David F. Labaree, *The Making of an American High School: The Credentials Market and the Central High School of Philadelphia, 1838–1939* (New Haven, CT: Yale University Press, 1988), 12; Persifor Frazier, "The Franklin Institute; Its Services and Deserts," *JFI* 165 (April 1908): 260.

130. Upper Darby Institute to George Escol Sellers, October 3, 1841, APS, Peale-Sellers Papers.

131. Franklin Institute, Minutes of the Board of Managers (1823–1831), Franklin Institute; Coleman and George Escol Sellers business/legal papers (1834–45), HSP, Cadwalader Collection, Series 6, box 103, folder 4.

132. Dawson, *Lives of the Philadelphia Engineers*, chap. 4, p. 7; Sellers, "Recollections of Coleman Sellers," 46.

133. Coleman Sellers, "President's Address, 1886," *Transactions of the American Society of Mechanical Engineers* 8 (1887): 691.

134. Coleman Sellers, "The Progress of the Mechanical Arts in Three Quarters of a Century," *JFI* 149 (January 1900): 5–25 (quote from 9).

135. Sellers, "President's Address," 681.

136. Diary of John Sellers, various entries, 1830–32, APS, Peale-Sellers Papers.

137. Ibid., 1833–39; see also Ashmead, *History of Delaware County*, 534, 549; *Report of the Proceedings of the First Meeting of the Sellers Family Association* (Philadelphia: Leeds & Biddle, 1910), 28.

138. Ashmead, *History of Delaware County*, 549.

139. Charles Sellers, Notes on Nathaniel Chapman's Medical Lectures (1828–29), HSP.

140. Coleman Sellers, Diary, November 1842–March 24, 1843, APS, Sellers Family Papers.

141. Ibid.

2. Migration Strategies and Industrial Frontiers

1. Eugene Ferguson, ed., *Early Engineering Reminiscences (1815–1840) of George Escol Sellers* (Washington, DC: Smithsonian Institution, 1965), 46.

2. David Jeremy, *Transatlantic Industrial Revolution: The Diffusion of Textile Technologies between Britain and America* (Cambridge, MA: MIT Press, 1981). On the formation of

social networks among industrial workers, see also Walter Licht, *Working for the Railroad: The Organization of Work in the Nineteenth Century* (Princeton, NJ: Princeton University Press, 1983); Philip Scranton, *Endless Novelty: Specialty Production and American Industrialization, 1865–1925* (Princeton, NJ: Princeton University Press, 1997); and Anthony F. C. Wallace, *Rockdale: The Growth of an American Village in the Early Industrial Revolution* (New York: Alfred A. Knopf, 1978). On the effects of transience upon individual cities and the broader integration of systems of cities, see Michael B. Katz, Michael J. Doucet, and Mark J. Stern, *The Social Organization of Early Industrial Capitalism* (Cambridge, MA: Harvard University Press, 1982), chap. 3.

3. Wallace, *Rockdale*, 219.

4. George Escol Sellers, *Memoirs*, vol. 4, 55–60, APS, Peale-Sellers Papers.

5. Ibid.

6. Ibid.

7. Ibid.

8. Ferguson, *Early Engineering Reminiscences*, 110.

9. Ibid., 108–34; John W. Maxson Jr., "George Escol Sellers: Inventor, Historian, and Papermaker," *Paper Maker* 38, no. 1 (1969): 40–57.

10. David Meyer, *The Roots of American Industrialization* (Baltimore: Johns Hopkins University Press, 2003), 70, 264; Anna T. Lincoln, *Wilmington, Delaware, Three Centuries under Four Flags, 1609–1937* (Port Washington, NY: Kennikat, 1972), 261; Thomas Scharf, *History of Delaware, 1609–1888* (Philadelphia: Richards, 1888), 797. On the growth of industrial Wilmington, see also Carol E. Hoffecker, *Wilmington, Delaware: Portrait of an Industrial City, 1830–1910* (Charlottesville: University Press of Virginia and Eleutherian Mills-Hagley Foundation, 1974).

11. Jeremy, *Transatlantic Industrial Revolution*.

12. Wallace, *Rockdale*, 118, 157.

13. Ibid.; Meyer, *Roots of American Industrialization*, 264.

14. Meyer, *Roots of American Industrialization*, note 14; also, Louis McLane, *Documents Relative to the Manufactures in the United States Collected and Transmitted to the House of Representatives, 1832, by the Secretary of the Treasury*, House Document no. 308, 1st session, 22nd Congress (Washington, DC: Duff Green, 1833).

15. Sean Wilentz, *Chants Democratic: New York City and the Rise of the American Working Class, 1788–1850* (New York: Oxford University Press, 1984), 151; "Manufacturing Industry of the State of New York," *Hunt's Merchants' Magazine*, October 1846, 370–72.

16. J. Morton Poole to Joseph Bancroft, September 19, 1830, Historical Society of Delaware (hereafter HSD), Poole Papers, J. Morton Poole Letters.

17. J. Morton Poole to Joseph Bancroft, August 6, 1832, HSD, Poole Papers, J. Morton Poole Letters.

18. Ibid.

19. Wallace, *Rockdale*, 225.

20. Scharf, *History of Delaware*, 781.

21. Peter Temin, "The Industrialization of New England, 1830–1880," in *Engines of Enterprise: An Economic History of New England*, ed. Peter Temin (Cambridge, MA: Harvard University Press, 2000); John S. Gilkeson Jr., *Middle-Class Providence, 1820–1940* (Princeton, NJ: Princeton University Press, 1986); Naomi Lamoreaux, *Insider Lending: Banks, Personal Connections, and Economic Development in Industrial New England* (New York: Cambridge University Press and NBER, 1994).

22. Louis C. Hunter, *A History of Industrial Power in the United States, 1780–1930*, vol. 1, *Waterpower in the Century of the Steam Engine* (Charlottesville: University Press of Virginia, 1979), 191; Louis C. Hunter, *A History of Industrial Power in the United States, 1780–1930*, vol. 2, *Steam Power* (Charlottesville: University Press of Virginia, 1985), 93–94.

23. See, for example, J. Morton Poole to Sarah Poole Bancroft and Joseph Bancroft, March 22, 1835, HSD, Poole Papers, J. Morton Poole Letters.

24. Albert Gallatin, *Considerations on the Currency and Banking System of the United States* (Philadelphia: Carey & Lea, 1831), 97–100.

25. J. Morton Poole to Sarah Poole Bancroft and Joseph Bancroft, March 22, 1835, HSD, Poole Papers, J. Morton Poole Letters.

26. J. Morton Poole to Sarah Poole Bancroft and Joseph Bancroft, March 22, 1835, HSD, Poole Papers, J. Morton Poole Letters.

27. On industrialization and the Second Great Awakening in Delaware County, see Wallace, *Rockdale*.

28. George Escol Sellers, *Memoirs*, vol. 4, 55–60, APS, Peale-Sellers Papers.

29. Wallace, *Rockdale*, 213.

30. J. Morton Poole to Samuel Poole, July 24, 1836, HSD, Poole Papers, J. Morton Poole Letters.

31. Wallace, *Rockdale*, 225.

32. J. Morton Poole to Joseph Bancroft, February 12, 1837, HSD, Poole Papers, J. Morton Poole Letters.

33. J. Morton Poole to Joseph Bancroft, July 9, 1838, HSD, Poole Papers, J. Morton Poole Letters.

34. Scharf, *History of Delaware*, 780.

35. J. Morton Poole to Sarah Poole Bancroft, August 2, 1835, HSD, Poole Papers, J. Morton Poole Letters.

36. "Our American Mechanical Engineers. No. I. William Sellers," *American Machinist* 2, no. 8 (June 28, 1879): 7.

37. William Poole Bancroft, Reminiscences, p. 7, HSD, Bird-Bancroft Collection.

38. George H. Sellers to William Poole Bancroft, January 20, 1846, HSD, Bird-Bancroft Collection, box 58, folder 8; Edward Bancroft to George H. Corliss, February 20, 1844, John Hay Library, Brown University, Corliss Papers; Monte Calvert, *The Mechanical Engineer in America, 1830–1910* (Baltimore: Johns Hopkins University Press, 1967), 8–11.

39. George Escol Sellers, *Memoirs*, vol. 5, 20, APS, Peale-Sellers Papers.

40. This argument is made most forcefully in William Cronon, *Nature's Metropolis: Chicago and the Great West* (New York: W. W. Norton, 1991).

41. George Escol Sellers, *Memoirs*, vol. 5, 22, APS, Peale-Sellers Papers.

42. For analysis of Cincinnati's competition with other gateways, see Cronon, *Nature's Metropolis*.

43. David Meyer, "A Dynamic Model of the Integration of Frontier Urban Places into the United States System of Cities," *Economic Geography* 56 (April 1980): 120–40; Charles Cist, *The Cincinnati Miscellany* (Cincinnati: Clark, 1845), 70.

44. Walter Stix Glazer, *Cincinnati in 1840: The Social and Functional Organization of an Urban Community during the Pre–Civil War Period* (Columbus: Ohio State University Press, 1999), 27–32; Daniel Aaron, *Cincinnati, Queen City of the West, 1819–1838* (Columbus: Ohio State University Press, 1992), 34–36, 133.

45. Charles Sellers to George Escol Sellers, March 4, 1844, APS, Peale-Sellers Papers.

46. Charles Sellers to Anna Sellers, November 12, 1843, APS, Sellers Family Papers.

47. Ibid.

48. Charles Sellers to Sophonisba Sellers, September 28, 1845, APS, Sellers Family Papers.

49. Ibid.

50. Ibid. See also Coleman Sellers to Sophonisba Sellers, April 19, 1846, APS, Sellers Family Papers.

51. Coleman Sellers to Sophonisba Sellers, April 19, 1846.

52. Charles Sellers to George Escol Sellers, November 13, 1845, APS, Sellers Family Papers.

53. Ibid.

54. *Chronicle of the Union League of Philadelphia, 1862–1902* (Philadelphia: Union League, 1902), 390–91.

55. Coleman Sellers, Diary, July 22, 1849, APS, Sellers Family Papers.

56. Sophonisba Sellers to Hannah Hill, July 29, 1846, APS, Sellers Family Papers.

57. Ibid.

58. Ibid.

59. Ibid.

60. Ibid.

61. Ibid.; see also, Coleman Sellers to Ann Sellers, March 15, 1846, APS, Sellers Family Papers.

62. Anna Sellers to Ann Sellers, April 1848; see also Charles Sellers to Coleman Sellers, April 23, 1848, both Peale-Sellers Papers, APS.

63. Coleman Sellers, Diary, August 2, 1849.

64. Charles Greve, *Centennial History of Cincinnati Representative Citizens* (Chicago: Biographical Publishing Co., 1904), 685.

65. Meyer, "Dynamic Model," 120–40; Cist, *Cincinnati Miscellany*, 70; William and Aimee Lee Cheek, "John Mercer Langston and the Cincinnati Riot of 1841," in *Race and the City: Work, Community, and Protest in Cincinnati, 1820–1970*, ed. Henry Louis Taylor Jr. (Chicago: University of Illinois Press, 1993), 29–69 (quote from 31).

66. Coleman Sellers, Diary, April 3, 1853; see also March 16, 1853.

67. On the diverse visions and models of schooling, see Michael B. Katz, *Reconstructing American Education* (Cambridge, MA: Harvard University Press, 1987).

68. Charles Sellers to Sophonisba Sellers, September 28, 1845, APS, Sellers Family Papers.

69. Coleman Sellers, Diary, January 25, 1854.

70. George Escol Sellers, *Improvements in Locomotive Engines, and Railways* (Cincinnati: Gazette Office, 1849), 3.

71. John H. White, *Cincinnati Locomotive Builders, 1845–1868* (Washington, D.C.: Smithsonian Institution, 1965), 58–63; Sellers, *Improvements in Locomotive Engines*, 19; Anna Sellers to Ann Sellers, April 1848, APS, Sellers Family Papers.

72. White, *Cincinnati Locomotive Builders*, 58–63; Sellers, *Improvements in Locomotive Engines*, 19.

73. Sellers, *Improvements in Locomotive Engines*; White, *Cincinnati Locomotive Builders*, 58–63.

74. Ferguson, *Early Engineering Reminiscences*, xvii.

75. Charles Sellers & Co., liquidation contract (September 28, 1846), APS, Peale-Sellers Papers.

76. George Escol Sellers, Plan for Establishing a Rail-road to San Francisco (c. 1849), APS, Peale-Sellers Papers.

77. Ibid.

78. Ibid.

79. White, *Cincinnati Locomotive Builders*, 65–84; see also remarks of Coleman Sellers, *Proceedings of the Franklin Institute* (October 18, 1865), in *Journal of the Franklin Institute* 80 (October 1865): 353. The Panama Railroad's minute books (1848–55) are at the National Archives, Washington, DC.

80. White, *Cincinnati Locomotive Builders*, 65–84; see also remarks of Sellers, *Proceedings of the Franklin Institute*; Coleman Sellers to George Escol Sellers, August 1, 1850, APS, Sellers Family Papers.

81. White, *Cincinnati Locomotive Builders*, 65–84.

82. Sophonisba Sellers to Ann Sellers, July 14, 1848, APS, Sellers Family Papers.

83. Ibid.

84. Ashmead, *History of Delaware County*, 549.

85. Elizabeth Sellers to Darby Monthly Meeting, March 11, 1845, FHL.

86. *Constitution of the Free Produce Society of Pennsylvania* (Philadelphia: D&S Neall, 1827); William Still, *The Underground Railroad*, rev. ed., vol. 2 (Middlesex, UK: Echo Library, 2006), 364–65.

87. Ashmead, *History of Delaware County*, 549.

88. Duff Green to Sellers, February 20, 1849; see also George Escol Sellers to Charles Sellers, September 2, 1848; George Escol Sellers to Duff Green, February 16 and March 5, 1849; Duff Green to George Escol Sellers, February 27, 1849, all APS, Peale-Sellers Papers.

89. Fletcher Green, "Duff Green: Industrial Promoter," *Journal of Southern History* 2, no. 1 (February 1936): 29–42.

90. Duff Green to George Escol Sellers, February 20, 1849.

91. Licht, *Working for the Railroad*.

92. Coleman Sellers to Aubrey H. Smith, March 9, 1850, APS, Sellers Family Papers.

93. Duff Green to George Escol Sellers, February 20, 1849; see also George Escol Sellers to Duff Green, February 21, 1849, both APS, Peale-Sellers Papers.

94. Duff Green to George Escol Sellers, October 19, 1849, APS, Peale-Sellers Papers.

95. Coleman Sellers to Aubrey H. Smith, March 9, 1850, APS, Sellers Family Papers.

96. Charles Sellers to Coleman Sellers, December 4, 1849, APS, Sellers Family Papers.

97. Coleman Sellers to George Escol Sellers, June 10, 1850, APS, Sellers Family Papers.

98. Charles Sellers to George Escol Sellers, October 7, 1850, APS, Sellers Family Papers.

99. Sophonisba Sellers to George Escol Sellers, June 1, 1851, APS, Sellers Family Papers.

100. White, *Cincinnati Locomotive Builders*, 65–84; remarks of Coleman Sellers, *Proceedings of the Franklin Institute*.

101. Coleman Sellers, Diary, January 1, 1853–June 14, 1854; Coleman Sellers to Sophonisba Sellers, March 20, 1853, both APS, Sellers Family Papers; also, White, *Cincinnati Locomotive Builders*, 94–98.

102. Coleman Sellers, Diary, January 29, 1854, APS, Sellers Family Papers.

103. Coleman Sellers, "The Progress of the Mechanical Arts in Three Quarters of a Century," *JFI* 149 (January 1900): 10.

104. *Railroad Advocate* 1 (November 18, 1854): 1.

105. Coleman Sellers, Diary, March 25, 1854, APS, Sellers Family Papers.

106. Coleman Sellers to Sophonisba Sellers, December 23, 1855, APS, Peale-Sellers Papers.

107. R. G. Dun, Ohio, vol. 78, p. 386 (January 26, 1856), Baker Library, Harvard Business School.

108. On Ohio's rail boom and bust, see White, *Cincinnati Locomotive Builders*, 5–9.

109. Bellefontaine & Indiana Railroad Co., *Third Annual Report of the President and Directors . . . to the Stockholders* (Cleveland: Sanford & Hayward, 1853), 6–7, 10.

110. Ibid., 11.

111. Ibid.

112. Charles Sellers to Anna Sellers, November 11, 1859, APS, Sellers Family Papers.

113. Ferguson, *Early Engineering Reminiscences*, xvii.

114. Harold Sellers Colton, "Mark Twain's Literary Dilemma and Its Sequel," *Arizona Quarterly* 17 (Autumn 1961): 229–32.

115. Ibid., 131–32.

116. Ibid.

117. Ferguson, *Early Engineering Reminiscences*; Charles Sellers to George Escol Sellers, September 20, 1863, APS, Peale-Sellers Papers.

118. Mark Twain, *The American Claimant* (New York: Oxford University Press, 1996).

119. "Death of an Old Engineer: George Sellers Dies at His Home on Mission Ridge," *Atlanta Constitution*, January 2, 1899, 3.

3. Rationalizing the Factory and City

1. William Sellers & Co. to Coleman Sellers, December 22, 1855, APS, Sellers Family Papers.

2. John K. Brown, "When Machines Became Gray and Drawings Black and White: William Sellers and the Rationalization of Mechanical Engineering," *IA: The Journal of the Society for Industrial Archeology* 25, no. 2 (1999): 29–54, quote on 32, quoting a patent extension the firm filed in 1872, from the National Archives, College Park, MD.

3. Sellers & Co. and other firms are personified as "they," as they were in the nineteenth century, when people associated companies named for their owners quite literally with the proprietors themselves.

4. Rationalization at William Sellers & Co. is recounted in Brown, "When Machines Became Gray and Drawings Black and White," to which this chapter traces much of its argument about rationalization's impacts on factory production.

5. *Report of the Twenty-Second Exhibition of American Manufactures Held in the City of Philadelphia, from the 19th to the 30th of October, Inclusive, 1852 by the Franklin Institute* (Philadelphia, 1852), 14.

6. See, for example, Edwin T. Freedley, *Philadelphia and Its Manufactures: A Hand-book Exhibiting the Development, Variety, and Statistics of the Manufacturing Industry of Philadelphia in 1857* (Philadelphia: Young, 1859).

7. On the importance of industrial districts for organizing urban and economic development, see Robert Lewis, *Chicago Made: Factory Networks in the Industrial Metropolis* (Chicago: University of Chicago Press, 2008).

8. Coleman Sellers, *American Supremacy in Applied Mechanics: A Series of Papers Reprinted from the Engineering Magazine* (New York: Engineering Magazine Co., 1892), 13–14, APS, Sellers Family Papers.

9. Alfred D. Chandler Jr., *The Visible Hand: The Managerial Revolution in American Business* (Cambridge, MA: Harvard University Press, 1977), 77; Nathan Rosenberg, "Technological Change in the Machine Tool Industry, 1840–1910," *Journal of Economic History* 23, no. 4 (1963): 414–46.

10. William Sellers & Co., *A Treatise on Machine-Tools* (Philadelphia: J. B. Lippincott, 1877), 184.

11. R. G. Dun & Co., credit report on Bancroft & Sellers, April 16, 1852, PA vol. 133, p. 428, Historical Collections, Baker Library, Harvard Business School (microfilm copy at Hagley Museum and Library).

12. Charles Robson, *The Manufactories and Manufacturers of Pennsylvania of the Nineteenth Century* (Philadelphia: Galaxy, 1875), 17.

13. Joseph Wickham Roe, *English and American Tool Builders* (New Haven, CT: Yale University Press, 1916), 247.

14. Andrew Ure, *The Philosophy of Manufactures or an Exposition of the Scientific, Moral, and Commercial Economy of the Factory System in Great Britain* (London, 1835), 32–35.

15. "The only alternative to power distribution and subdivision within the factory was to provide each machine, each industrial operation or process, with its own source of power; that is to say, with its own waterwheel or steam engine." Louis C. Hunter, *A History of Industrial Power in the United States, 1780–1930*, vol. 1, *Waterpower in the Century of the Steam Engine* (Charlottesville: University Press of Virginia, 1979), 418.

16. William Sellers & Co., *Treatise on Machine-Tools*, 184.

17. Bancroft & Sellers order book (1848–54), Hagley, Soda House, Accession 1466, Item 1.

18. Ibid.

19. Ibid.

20. Almost a decade before his return to Philadelphia, Bancroft submitted two inventions for review by the Committee on Science and the Arts, a mechanical press to strengthen iron shafts and a steam engine governor that regulated the flow of water to engines. Committee on Science and the Arts, reports 220 (1839) and 232 (1840), Franklin Institute.

21. *Report of the Twentieth Exhibition of American Manufactures Held in the City of Philadelphia, from the 15th to the 26th of October, Inclusive, 1850 by the Franklin Institute* (Philadelphia, 1850), 2.

22. *Catalogue of the Twentieth Exhibition of American Manufactures Held in the City of Philadelphia, by the Franklin Institute . . . from the 14th day of Nov., to the 2d Day of Dec., 1854* (Philadelphia: Young, 1854), 7; Alexander E. Outerbridge Jr. and Coleman Sellers Jr., "William Sellers," *JFI* 159 (May 1905): 380.

23. *American Mechanics' Magazine*, vol. 1, no. 19 (June 11, 1825): 300–301; Hunter, *History of Industrial Power in the United States*, 1:463. See also R. G. Dun & Co., credit report on Sellers & Pennock, July 9, 1856–January 4, 1871, PA vol. 3, p. 88, Historical Collections, Baker Library, Harvard Business School (microfilm copy at Hagley Museum and Library).

24. Coleman Sellers to Edward Bancroft, March 19, 1852, APS, Sellers Family Papers.

25. Bancroft & Sellers order book, 22.

26. Thomas R. Heinrich, *Ships for the Seven Seas: Philadelphia Shipbuilding in the Age of Industrial Capitalism* (Baltimore: Johns Hopkins University Press, 1997), 18.

27. James Dredge, *The Pennsylvania Railroad: Its Organization, Construction, and Management* (New York: Wiley & Sons, 1879), Hagley Library.

28. Jeffrey P. Roberts, "Railroads and the Downtown: Philadelphia, 1830–1900," in *The Divided Metropolis: Social and Spatial Dimensions of Philadelphia, 1800–1975*, ed. William W. Cutler III and Howard Gillette Jr. (Westport, CT: Greenwood Press, 1980), 27–55.

29. Philip Scranton, *Endless Novelty: Specialty Production and American Industrialization, 1865–1925* (Princeton, NJ: Princeton University Press, 1997), 18–19; Robert Lewis, *Manufacturing Montreal: The Making of an Industrial Landscape, 1850 to 1930* (Baltimore: Johns Hopkins University Press, 2000).

30. John K. Brown, *The Baldwin Locomotive Works, 1831–1915: A Study in American Industrial Practice* (Baltimore: Johns Hopkins University Press, 1995), 7–8, 37, 135.

31. Deed Book TH 83, pp. 515–31, City Archives of Philadelphia.

32. Deed Books LRB 141, pp. 51, 208; LRB 227, p. 56; JTO 15, pp. 75–103; JTO 28, p. 485; all City Archives of Philadelphia.

33. Deed Book TH 132, p.170, City Archives of Philadelphia.

34. "Our American Mechanical Engineers. No. I. William Sellers," *American Machinist* 2, no. 8 (June 28, 1879): 7.

35. Robson, *Manufactories and Manufacturers of Pennsylvania*, 15.

36. I. L. Vansant, ed., *The Royal Road to Wealth, an Illustrated History of the Successful Business Houses of Philadelphia* (Philadelphia: Loag, c. 1868), 8.

37. Robson, *Manufactories and Manufacturers of Pennsylvania*, 15. Descriptions of the plant are also found in Edwin T. Freedley, ed., *Leading Pursuits and Leading Men: A Treatise on the Principal Trades and Manufactures of the United States* (Philadelphia: Young, 1856), 333; Edwin T. Freedley and J. Leander Bishop, *A History of American Manufactures from 1608 to 1860*, vol. 3 (Philadelphia: Young, 1868), 31.

38. William Allen, *Plan of the City of Philadelphia and Adjoining Districts* (Philadelphia: H. S. Tanner, 1830); *Plan of the City of Philadelphia and Adjoining Districts Enlarged with Numerous Additions and Corrections by F. J. Roberts* (Philadelphia, 1838).

39. See, for example, Sam Bass Warner, *The Private City: Philadelphia in Three Periods of Its Growth* (Philadelphia: University of Pennsylvania Press, 1968).

40. Eli K. Price, "Preface to the Digest of 1833," in District of Spring Garden Board of Commissioners, *A Digest of the Acts of Assembly, and of the Ordinances of the Inhabitants and Commissioners of the District of Spring Garden*, by Robert Bethell (Philadelphia: J. H. Jones, 1852), City Archives of Philadelphia.

41. District of Spring Garden Board of Commissioners, *Digest of the Acts of Assembly*.

42. Ibid. For a general discussion of the regulatory state in the nineteenth century, see William Novak, *The People's Welfare: Law and Regulation in Nineteenth-Century America* (Chapel Hill: University of North Carolina Press, 1996).

43. *Niles' Weekly Register* 17 (January 15, 1820): 335.

44. Kuhn's real estate holdings are characterized based on analysis of post–Civil War fire insurance atlases, especially Elvino Smith, *Atlas of the City of Philadelphia* (Philadelphia,

1875). Although he died in 1860, his estate continued to own many properties acquired earlier.

45. The seminal work on the political economy of growth machines and urban regimes is John Logan and Harvey Molotch, *Urban Fortunes: The Political Economy of Place* (Berkeley: University of California Press, 1988).

46. For an analysis of the political economy of consolidation, see Andrew Heath, *"The Manifest Destiny of Philadelphia": Imperialism, Republicanism, and the Remaking of a City and Its People, 1837–1877* (dissertation, University of Pennsylvania, 2008).

47. Eli K. Price, *The History of the Consolidation of the City of Philadelphia* (Philadelphia: J. B. Lippincott, 1873), 12.

48. Heath, *"Manifest Destiny of Philadelphia."* See also Howard Gillette Jr., "The Emergence of the Modern Metropolis: Philadelphia in the Age of Its Consolidation," in Cutler and Gillette, *Divided Metropolis*, 3–25; Sam Bass Warner, *The Private City: Philadelphia in Three Periods of Its Growth* (Philadelphia: University of Pennsylvania Press, 1968), 152; Michael Feldberg, *The Philadelphia Riots: A Study of Ethnic Conflict* (Westport, CT: Greenwood Press, 1975); Robert L. Bloom, "Morton McMichael's North American," *PMHB* 77 (April 1953); Andrew Dawson, *Lives of the Philadelphia Engineers: Capital, Class and Revolution, 1830–1890* (Burlington, VT: Ashgate, 2004).

49. *Price List and Commercial Advertiser*, February 18, 1854.

50. Freedley, *Philadelphia and Its Manufactures*, 314.

51. Ibid., 316.

52. Ibid, 425.

53. R. G. Dun & Co., credit report on Bancroft & Sellers, July 25, 1855, and April 23, 1864, PA vol. 133, pp. 428, 547.

54. The 1870 census would reveal that the city accounted for nearly one-third of total U.S. output, while the whole of New England produced 40 percent. Ross M. Robertson, "Changing Production of Metalworking Machinery, 1860–1920," in *Output, Employment and Productivity after 1800* (New York: NBER, 1966), 489.

55. Deed Book RDW 6, p. 286, City Archives of Philadelphia; R. G. Dun & Co., credit report on Bancroft & Sellers, January 8, 1854, and October 3, 1855, PA vol. 133, p. 428.

56. Brown, "When Machines Became Gray," 49 n. 7. For historian of technology Lindy Biggs, by contrast, "rationalization in industry . . . refers to the introduction of predictability and order—machinelike order—that eliminates all questions of how work is to be done, who will do it, and when it will be done." This definition of rationalization as a rigid process more accurately fits the mass-producing industries of the early twentieth century—the main focus of Biggs's work—but does not aptly describe the more flexible approach of specialty manufacturers like Sellers & Co. Lindy Biggs, *The Rational Factory: Architecture, Technology, and Work in America's Age of Mass Production* (Baltimore: Johns Hopkins University Press, 1996), 6. See also Alfred Chandler, *Strategy and Structure* (Cambridge, MA: MIT Press, 1962), 388; David F. Noble, *Forces of Production: A History of Production* (New York: Alfred A. Knopf, 1984), 267.

57. Sellers & Co., *Treatise on Machine-Tools*, 218.

58. Brown, "When Machines Became Gray," 31.

59. *American Machinist* 1, no. 5 (March 1878): 2.

60. Vansant, *Royal Road to Wealth*, 15.

61. *American Machinist* 1, no. 5 (March 1878): 2. See also William Sellers & Co., Estimates for Shafting (1867–72), Hagley, Soda House; Sellers, "Transmission of Motion," *JFI* 94 (1872): 317; Robson, *Manufactories and Manufacturers of Pennsylvania*, 16; Hunter, *History of Industrial Power in the United States*, 1:475; Brown, "When Machines Became Gray," 31.

62. "Our American Mechanical Engineers," 7.

63. Brown, "When Machines Became Gray," 46.

64. Testimony of William Sellers & Co. manager Justus William Schwacke, *Industrial Relations: Final Report and Testimony Submitted to Congress by the Commission on Industrial Relations*, vol. 3 (Washington, DC: Government Printing Office, 1916), 2894–95.

65. Robertson, "Changing Production of Metalworking Machinery," 485–86.

66. Brown, "When Machines Became Gray," 32, 45.

67. Ibid.; see also Henry Roland, "Six Examples of Successful Shop Management," *Engineering Magazine* 12 (February 1897): 831–37.

68. Joseph D. Weeks, "Report on the Statistics of Wages in Manufacturing Industries," in U.S. Department of the Interior, Census Office, *Tenth Census of the United States* (Washington, DC, 1886), 20: 167–533.

69. William Sellers & Co. Scrapbook (1882–1922), Pennsylvania State Archives; Brown, "When Machines Became Gray," 32.

70. William Sellers & Co. drawings survive in the collections of the Franklin Institute, the Pennsylvania State Archives, and the Rochester, New York, firm Machine Tool Research. They provide an in-depth view into the company's design and production practices, which is masterfully treated in Brown, "When Machines Became Gray." See also Coleman Sellers Jr., "Mechanical Drawing: A Lecture Delivered before the Franklin Institute, November 11th, 1881," *JFI* 113 (January 1882): 32–49; J. Sellers Bancroft, discussion of Frederick O. Ball, "Drafting-Room and Shop Systems," *Transactions of the American Society of Mechanical Engineers* 22 (1901): 1040–72.

71. Brown, "When Machines Became Gray," 33.

72. Bancroft, discussion of Ball, 1067.

73. Ibid., 44; see also John K. Brown, "Design Plans, Working Drawings, National Styles: Engineering Practice in Great Britain and the United States, 1775–1945," *Technology and Culture* 41, no. 2 (April 2000): 208.

74. Freedley, *Leading Pursuits and Leading Men*, 332; Brown, "Design Plans," 208.

75. *McElroy's Philadelphia City Directory* (Philadelphia: McElroy, 1860).

76. Brown, "When Machines Became Gray," 39–40, 52 n. 62.

77. Coleman Sellers, "An Old Photographic Club," reprinted from *Anthony's Photographic Bulletin* (1888), 2, APS, Sellers Family Papers; see also "In Memoriam: Coleman Sellers," *JFI* 165 (March 1908): 165–72.

78. "A Pennsylvanian: Coleman Sellers," *Evening Bulletin*, July 21, 1924, 14, Sellers Library, Scrapbooks, vol. 15.

79. Coleman Sellers, Patent Abstract, U.S. Patent 31,357 (patented February 5, 1861), reprinted in David Rogers, *Inventions and Their Inventors* (Hertford, UK: M-Y Books, 2010), 11.

80. Sellers, "An Old Photographic Club"; Coleman Sellers letter books (1863–76); Coleman Sellers to George Escol Sellers (January 2, 1865), all APS, Sellers Family Papers.

81. Letters between Coleman and Holmes are in APS, Sellers Family Papers.

82. R. Eskind, "Coleman Sellers; Engineer, Photographer, Correspondent," typescript copy of a speech given at "Early Photography in Pennsylvania" (Penn State symposium, March 31, 1984), William Sellers & Co. Records, Pennsylvania State Archives.

83. Sellers, "Old Photographic Club," 14.

84. Ibid.

85. Eskind, "Coleman Sellers."

86. Roe, *English and American Tool Builders*, 247; Brown, "When Machines Became Gray," 50 n. 29.

87. Outerbridge and Sellers, "William Sellers," 372–73.

88. William Sellers & Co., *Treatise on Machine-Tools*, 267.

89. William Sellers & Co. to Enoch Lewis, October 15, 1861, Enoch Lewis Papers, Andrew Carnegie Papers, Library of Congress. I am indebted to John K. Brown for this source.

90. James A. Mulholland to Coleman Sellers, May 17, 1864, APS, Sellers Family Papers.

91. Freedley and Bishop, *History of American Manufactures*, 30.

92. Dredge, *Pennsylvania Railroad*, 84.

93. "Our American Mechanical Engineers," 7.

94. Outerbridge and Sellers, "William Sellers," 373.

95. William Sellers & Co. Scrapbook, Pennsylvania State Archives; Brown, "When Machines Became Gray," 50 n. 29.

96. Sellers, *American Supremacy in Applied Mechanics*, 13; William Sellers & Co., *Treatise on Machine-Tools*, 147–48, 267; Freedley and Bishop, *History of American Manufactures*, 30–31; Wm. Sellers & Co., "Hydraulic Riveting Machines," *JFI* 102, no. 5 (November 1876): 305–12.

97. William Sellers & Co., *Treatise on Machine-Tools*, 64, 101, 103, 135, 151, 163–64, 174, 178, 182, 268.

98. The history of interchangeable parts manufacturing in America is reviewed in David Hounshell, *From the American System to Mass Production, 1800–1932: The Development of Manufacturing Technology in the United States* (Baltimore: Johns Hopkins University Press, 1984); see also Merritt Roe Smith, *Harpers Ferry Armory and the New Technology: The Challenge of Change* (Ithaca, NY: Cornell University Press, 1977). While Hounshell, Smith, and other historians of technology argue that the development of interchangeability was slow, limited, and fraught with technical problems in the manufacture of everything from guns to sewing machines to mechanical reapers, they pay relatively little attention to the machine tool sector, where the systems and tools that enabled interchangeability were largely fashioned. And although they focus on the challenges facing the perfection of interchangeable parts manufacturing, Hounshell and Smith ultimately demonstrate the central importance of these systems for American industry and material culture. See also Rosenberg, "Technological Change in the Machine Tool Industry."

99. William Sellers, "Machinery Manufacturing Interests," in *One Hundred Years of American Commerce*, vol. 2, ed. Chauncey M. Depew (New York: Haynes, 1895), 348.

100. William Sellers & Co., *Treatise on Machine-Tools*, 11.

101. Robson, *Manufactories and Manufacturers of Pennsylvania*, 15.

102. William Sellers & Co., *Treatise on Machine-Tools*, 113.

103. Freedley and Bishop, *History of American Manufactures*, 28. See also Outerbridge and Sellers, "William Sellers," 369.

104. See for example the following, all by William Sellers & Co.: *A Treatise on Shafting* (Philadelphia, 1883); *A Treatise on Shaping, Slotting, and Planing Machines* (Philadelphia, 1884); *A Treatise on Shaping, Slotting, and Planing Machines* (Philadelphia, 1887); *Illustrated Catalogue and General Description of Improved Machine Tools for Working Metal* (Philadelphia: J. B. Lippincott, 1895); *Shafting* (Philadelphia, 1895); *Vicars' Patent Automatic Mechanical Stoker* (Philadelphia, 1895); *Shafting* (Philadelphia, 1896); *Worm Geared Hoisting Machines* (Philadelphia, 1900); *Steam Riveting Machines* (Philadelphia, 1900); *Slotting Machines, Belt and Motor Driven* (Philadelphia, 1900); *Sellers Drill Grinding Machine* (Philadelphia, 1900); *Patent Improved Drill Grinding Machine for Grinding Twist Drills and Flat Drills* (Philadelphia, 1900); *Metal-Working Machine Tools for Railway and Machine-Shop Equipment* (Philadelphia, 1900); *Centrifugal Sand-Mixing Machine for Mixing All Kinds of Foundry Sands and Compounds* (Philadelphia, 1900); *Shafting* (Philadelphia, 1901); *Metal-Working Machine Tools for Railway and Machine-Shop Equipment* (Philadelphia, 1904); *A Treatise on Patent Improved Drill-Grinding Machines* (Philadelphia, 1907); *Machine Tools for Railway and Machine Shop Equipment: Shafting and Injectors* (Philadelphia, 1911); *Shafting and Power Transmission Material* (Philadelphia, 1916); *Locomotive Injectors and Boiler Attachments* (Philadelphia, 1922); *Sellers Precision-Built Machine Tools, since 1848* (Philadelphia, 1942).

105. William Sellers & Co., *Treatise on Machine-Tools*, 40.

106. Scranton, *Endless Novelty*, 62; Board of Managers Minutes, vol. 2 (1848–70), 134, Franklin Institute.

107. Board of Managers Minutes, vol. 2, 178–276; Scranton, *Endless Novelty*, 62.

108. Scranton, *Endless Novelty*, 63.

109. Board of Managers Minutes, vol. 2, 279–303.

110. Bruce Sinclair, *Philadelphia's Philosopher Mechanics: A History of the Franklin Institute, 1824–1865* (Baltimore: Johns Hopkins University Press, 1974), 311; Scranton, *Endless Novelty*, 63.

111. Board of Managers Minutes, vol. 2, 279.

112. See, for example, Sellers & Co., "Hydraulic Riveting Machines"; Coleman Sellers, "Theory and Construction of the Self-Acting Slide Lathe," *JFI* 64, no. 1 (July 1872): 106–12.

113. "Proceedings of the Stated Monthly Meeting, April 21, 1864," *JFI* 77 (May 1864): 344.

114. Ibid., 318–490.

115. William Sellers's campaign for his standard screw thread is recounted in Bruce Sinclair, "At the Turn of a Screw: William Sellers, the Franklin Institute, and a Standard American Thread," *Technology and Culture* 10, no. 1 (January 1969): 20–34; Roe, *English and American Tool Builders*, 248–49.

116. *Scientific American*, October 10, 1863, 233.

117. "Proceedings of the Stated Monthly Meeting, April 21, 1864," *JFI* 77 (May 1864): 344.

118. Bancroft & Sellers, order book.

119. Dredge, *Pennsylvania Railroad*, 77.

120. Sinclair, "At the Turn of a Screw," 28.

121. John S. Gilkeson Jr., *Middle-Class Providence, 1820–1940* (Princeton, NJ: Princeton University Press, 1986), 59.

122. "Proceedings of the Stated Monthly Meeting, April 21, 1864," 350–51.

123. Ibid., 351.

124. *Scientific American*, October 29, 1864, 278; *Scientific American*, March 4, 1865, 151.

125. Scranton, *Endless Novelty*, 64.

126. *JFI* 79 (January 1865): 56; Sinclair, "At the Turn of a Screw," 29.

127. *Report of the Board to Recommend a Standard Gauge for Bolts, Nuts and Screws Threads for the United States Navy, May 1868* (Washington, DC, 1880), 22–23, 30; Sinclair, "At the Turn of a Screw," 24–28.

128. *History and Early Reports of the Master Car-Builders' Association* (New York, 1885), 80.

129. Coleman Sellers, Address in *Seventh Annual Report of the American Railway Master Mechanics Association* (1874), 173.

130. *American Machinist* 1, no. 1 (November 1877): 3.

131. George R. Stetson, "Standard Sizes of Screw Threads," *A.S.M.E., Transactions* 1 (1880): 125; also, Sinclair, "At the Turn of a Screw," 32.

132. *Memorial of Alexander Lyman Holley* (New York: American Institute of Mining Engineers, 1884), 88, 123; Jeanne McHugh, *Alexander Holley and the Makers of Steel* (Baltimore: Johns Hopkins University Press, 1980), 64.

133. According to the National Institute of Standards and Technology, by the beginning of the twenty-first century there were close to eight hundred thousand global standards. James Surowiecki, "Turn of the Century: How One Man with a Vision and a Simple Screw Launched the Mass Production Era," *Wired*, January 2002, 85.

134. Scranton, *Endless Novelty*, 63; see also Dawson, *Lives of the Philadelphia Engineers*, 62–63.

135. Coleman Sellers, "The Metric System in Our Work-Shops; Will Its Value in Practice Be an Equivalent for the Cost of Its Introduction?" *JFI* 67 (June 1874): 381–91; Coleman Sellers and W. P. Tatham, "Report on the Metric System," *JFI* 101 (1876): 370–81; Coleman Sellers, "The Metric System—Is It Wise to Introduce It Into Our Machine Shops?" *JFI* 110 (November 1880): 289–312; Frederick Fraley, "An Essay on Mechanics and the Progress of Mechanical Science—1824 to 1882," *Transactions of the ASME* 3 (1882): 205. See also Monte Calvert, *The Mechanical Engineer in America, 1830–1910* (Baltimore: Johns Hopkins University Press, 1967), 179–86; Dawson, *Lives of the Philadelphia Engineers*, 62–63.

136. Coleman Sellers, "Standard Reamers for Locomotive Work," *JFI* 114 (September 1883): 161–70. He originally presented this paper at the 1883 Master Mechanics' Association convention in Chicago.

137. Sellers, "Mechanical Drawing," 32–49.

138. Sellers, "Machinery Manufacturing Interests," 348; Sinclair, "At the Turn of a Screw," 31.

139. *JFI* 79 (January 1865): 53–57; "The Sellers or Franklin Institute System of Screw-Threads," *JFI* 123 (April 1887): 261–76; *JFI* 139 (March 1895): 319–20; Coleman Sellers III, "The Sellers or Franklin Institute Standard Screw Thread," *JFI* 247 (February 1949): 95–99; Roe, *English and American Tool Builders*, 249; Sinclair, "At the Turn of a Screw."

4. Progressive Economic Development

1. "Officers and Members of the Saturday Club, 1877–1878," HSP, Edward Carey Gardiner Collection, box 60, folder 18.
2. Ibid.
3. Abram S. Hewitt to A. J. Drexel, November 10, 1881, HSP, Gratz Manuscripts, case 8, box 11; Chester Arthur to A. J. Drexel, January 17, 1881, HSP, Society Collection. For another example of national networking, see *Banquet Given by the Learned Societies of Philadelphia at the American Academy of Music September 17, 1887, Closing the Ceremonies in Commemoration of the Framing and Signing of the Constitution of the United States* (Philadelphia, 1888), Union League Archives.
4. Horace H. Furness, *F.R., 1833–1900* (Philadelphia, 1903), 14, in the collection of George E. Thomas.
5. The rise of the Republican machine is chronicled in Peter McCaffery, *When Bosses Ruled Philadelphia: The Emergence of the Republican Machine, 1867–1933* (University Park: Penn State University Press, 1993).
6. For an account of the transatlantic networks of people and institutions that shaped Progressivism, see Daniel Rodgers, *Atlantic Crossings: Social Politics in a Progressive Age* (Cambridge, MA: Harvard University Press, 1998).
7. On the significance of these networks for the industrial base of Philadelphia and other regions of specialty and batch manufacturing in the late nineteenth century, see Philip Scranton, *Endless Novelty: Specialty Production and American Industrialization, 1865–1925* (Princeton, NJ: Princeton University Press, 1997), 157–60.
8. For an overview of Philadelphia industry's wartime activities, see J. Matthew Gallman, *Mastering Wartime: A Social History of Philadelphia during the Civil War* (New York: Cambridge University Press, 1990).
9. General discussions of the Union League and manufacturers' involvement in the Republican Party include Maxwell Whiteman, *Gentlemen in Crisis: The First Century of the Union League of Philadelphia, 1862–1962* (Philadelphia: Union League, 1975); Andrew Dawson, *Lives of the Philadelphia Engineers: Capital, Class and Revolution, 1830–1890* (Burlington, VT: Ashgate, 2004).
10. Arthur M. Lee, "Henry C. Carey and the Republican Tariff," *PMHB* 81 (July 1957): 280–302; William Dusinberre, *Civil War Issues in Philadelphia, 1856–1865* (Philadelphia: University of Pennsylvania Press, 1965), 77–78; *Public Ledger*, May 5, 1858, 2; Thomas Scharf and Thompson Westcott, *History of Philadelphia, 1609–1884* (Philadelphia: J. H. Everts, 1884), 1:728. On Republican politics nationally, see Eric Foner, *Free Soil, Free Labor, Free Men: The Ideology of the Republican Party before the Civil War* (New York: Oxford University Press, 1970).
11. James E. Harvey, *Testimonial to Henry C. Carey, Esq.* (Philadelphia: Collins, 1859); Board of Trade, *Annual Report* (Philadelphia: Saturday Evening Post, 1858).
12. "The Original Constitution and List of Members of the First Republican Club, Formed in the City of Philadelphia, 1856" (Philadelphia, 1856), Union League Archive, Philadelphia. See also Dawson, *Lives of the Philadelphia Engineers*, chap. 4.
13. George W. Simpson to Coleman Sellers, January 20, 1860, APS, Sellers Family Papers.
14. George W. Simpson to Coleman Sellers, December 18, 1860, APS, Sellers Family Papers.
15. John K. Brown, correspondence with the author, May 15, 2004.

16. Dawson, *Lives of the Philadelphia Engineers*, 140–41. See also Scharf and Westcott, *History of Philadelphia*, 1:739–45; Dusinberre, *Civil War Issues*, 106; George P. Lathrop, *History of the Union League of Philadelphia* (Philadelphia: J. B. Lippincott, 1884), 12.

17. John Sellers's narration, in *A Memorial of the Union Club of Philadelphia* (Philadelphia: J. B. Lippincott, 1871), 391, Union League Archives.

18. Ibid.

19. Ibid.

20. Dawson, *Lives of the Philadelphia Engineers*, 142. See also Frank Taylor, *Philadelphia in the Civil War, 1861–1865* (Philadelphia: City of Philadelphia, 1913), 216, 246.

21. Union League Board Minutes, April 8, 1863, Union League Archives.

22. Union League of Philadelphia, *Address by the Union League of Philadelphia, to the Citizens of Pennsylvania, in Favor of the Re-Election of Abraham Lincoln* (Philadelphia, 1864), Union League Archives.

23. *The Visitor's Guide to the Great Central Fair, for the U.S. Sanitary Commission, Held in Logan Square, Philadelphia, June, 1864* (Philadelphia: Selheimer, 1864).

24. The invitation is in the Sanitary Fair folder of the Union League Archives. Also, Coleman Sellers to George Escol Sellers, April 15, 1864, and May 23, 1864, APS, Sellers Family Papers; Gallman, *Mastering Wartime*, 146.

25. Coleman Sellers to U.S. Commissioner of Patents, September 27, 1861; Coleman Sellers, "Memorandum re-invention" (1861), both APS, Sellers Family Papers.

26. J. J. S. Laidley to Brig. Gen. J. W. Ripley, June 16, 1863; S. V. Benet to Gen. A. B. Dyer, December 10, 1864, February 10, 1865, and June 20, 1867, all National Archives, Philadelphia, Record Group 156, Frankford Arsenal, Letters Sent to Chief of Ordnance by Commanding Officers.

27. Estimates for Shafting and Machine Tools, 1862–72, William Sellers & Co. Records, Item 3, Hagley, Soda House.

28. Coleman Sellers to George Escol Sellers, July 20, 1863; also, Coleman Sellers to George Escol Sellers, June 23, 1863, and Coleman Sellers to Henry Greenwood, June 21, 1863, all APS, Sellers Family Papers.

29. Dawson, *Lives of the Philadelphia Engineers*, chap. 1.

30. Gallman, *Mastering Wartime*, 258, table 10.2.

31. *The Rich Men of Philadelphia: Income Tax of the Residents of Philadelphia and Bucks County for the Year Ending April 30, 1865* (Philadelphia: John Trentwith, 1865), 5, University of Pennsylvania Library.

32. Wm. Sellers & Co. to Coleman Sellers, December 24, 1863; Isaac P. Morris to Coleman Sellers, December 17–18, 1864; Wm. Sellers & Co. to Coleman Sellers, January 19, 1866; Coleman Sellers to Wm. Sellers & Co., January 20, 1866, all APS, Sellers Family Papers.

33. R. G. Dun & Co., credit report on Bancroft & Sellers, April 23, 1864, and April 12, 1873, PA vol. 133, p. 547.

34. R. G. Dun & Co., credit report on William Sellers & Co., October 28, 1885, April 14 and October 5, 1886, PA vol. 134, p. 786.

35. *McElroy's Philadelphia Directory* (Philadelphia: Biddle, 1856), 252.

36. William Sellers to Amélie Sellers, August 1, 1883, Hagley, Soda House.

37. Amélie Sellers to William Sellers, August 7, 1883, Hagley, Soda House.

38. Amélie Sellers to William Sellers, August 11, 1883, Hagley, Soda House.

39. Sven Beckert, *The Monied Metropolis: New York City and the Consolidation of the American Bourgeoisie, 1850–1896* (New York: Cambridge University Press, 2001), 115. On the national ascendancy of local capitalists, see also E. Digby Baltzell, *Philadelphia Gentlemen: The Making of a National Upper Class* (New York: Free Press, 1958).

40. Coleman Sellers, *American Supremacy in Applied Mechanics: A Series of Papers Reprinted from the Engineering Magazine* (New York: Engineering Magazine Co., 1892), 47, APS, Sellers Family Papers.

41. Global economic integration in the late nineteenth century is discussed in Kevin O'Rourke and Jeffrey Williamson, *Globalization and History: The Evolution of a Nineteenth-Century Atlantic Economy* (Cambridge, MA: MIT Press, 1999); Michael Bordo, Alan M. Taylor, and Jeffrey Williamson, *Globalization in Historical Perspective* (Chicago: University of Chicago Press, 2003).

42. Beckert, *Monied Metropolis.* See also Iver Bernstein, *The New York City Draft Riots: Their Significance in American Society and Politics in the Age of the Civil War* (New York: Oxford University Press, 1990). Beckert and Bernstein see a split between mercantile and manufacturing capital in New York, and Beckert argues that the industrialists' interests trumped those of the merchant class during the Civil War, giving them the upper hand in New York's capitalist class. However, at least in the upper echelons of Philadelphia capital there appeared more genuinely mutual interests and collaboration.

43. *Chronicle of the Union League of Philadelphia, 1862–1902* (Philadelphia: Union League, 1902), 173–74.

44. John Sellers's narration, in *A Memorial of the Union Club of Philadelphia*, 391, Union League Archives.

45. William Sellers's narration, ibid., 383.

46. McCaffery, *When Bosses Ruled Philadelphia*, 9.

47. Whiteman, *Gentlemen in Crisis*, 114–15.

48. McCaffery, *When Bosses Ruled Philadelphia*, 51, appendixes 1 and 2.

49. William Sellers to Henry Charles Lea, January 29, 1880; November 8, November 30, December 5, and December 13, 1886; January 9, 1895, University of Pennsylvania, Annenberg Special Collections, Henry Charles Lea Papers. See also William Sellers to Herbert Welsh, January 2, 1895, and January 24, 1901, HSP, Herbert Welsh Collection.

50. William Sellers to Herbert Welsh, July 6, 1891, HSP, Herbert Welsh Collection.

51. Ibid.

52. Ibid.

53. Philadelphia Social Science Association, *Report of the Executive Committee, Read at the Annual Meeting, December 12, 1871* (Philadelphia: Penn Monthly, 1872), 3.

54. See, for example, Social Science Association of Philadelphia, *Papers of 1874* (Philadelphia, 1874); Edmund J. James, "The Relation of the Modern Municipality to the Gas Supply: Read at a Meeting of the Association, February 11, 1886" (Philadelphia Social Science Association, 1886); Simon Nelson Patten, "The Principles of Rational Taxation: Read at a Meeting of the Association, November 21st, 1889" (Philadelphia: PSSA, 1889).

55. "Annual Report of the Philadelphia Social Science Association" for 1873, in Social Science Association of Philadelphia, *Papers of 1874*, 35.

56. As the social sciences gained prominence within higher education, in 1890 the Philadelphia Social Science Association merged with the American Academy of Political and Social Science. Dawson, *Lives of the Philadelphia Engineers*, chap. 5; Andrew Dawson, "The Workshop and the Classroom: Philadelphia Engineering, the Decline of Apprenticeship, and the Rise of Industrial Training, 1878–1900," *History of Education Quarterly* 39, no. 2 (Summer 1999): 143–60; Joseph G. Rosengarten, "Work of the Philadelphia Social Science Association," *Annals of the American Academy of Political and Social Science* 1 (1890–91): 709–18; Thomas L. Haskell, *The Emergence of Professional Social Science: The American Social Science Association and the Nineteenth Century Crisis of Authority* (Urbana: University of Illinois Press, 1977), 113, 127; Mary Furner, *From Advocacy to Objectivity: A Crisis in Professionalization of American Social Science, 1865–1905* (Lexington: University Press of Kentucky, 1975).

57. Roland Falkner, "Editorial," *Annals of the American Academy of Political and Social Science* 7 (1896): 74–77; Winton Solberg, "Edmund Janes James Builds a Library: The University of Illinois Library, 1904–1920," *Libraries and Culture* 39, no. 1 (2004): 36–75.

58. For example, James, "Relation of the Modern Municipality to the Gas Supply."

59. Andrew Dawson, "Reassessing Henry Carey (1793–1879): The Problems of Writing Political Economy in Nineteenth-Century America," *Journal of American Studies* 34, no. 3 (December 2000): 465–85.

60. Dawson, *Lives of the Philadelphia Engineers*, 173. Also, Whiteman, *Gentlemen in Crisis*, 146–47.

61. Alexander E. Outerbridge Jr. and Coleman Sellers Jr., "William Sellers," *JFI* 159 (May 1905): 367.

62. Franklin Institute, Board of Managers Minutes, vol. 3 (1870–86).

63. Dawson, *Lives of the Philadelphia Engineers*, 167.

64. Franklin Institute, Board of Managers Minutes, vol. 2 (1848–70), 487–97; also, Whiteman, *Gentlemen in Crisis*, 132.

65. *Report of the Twenty-Seventh Exhibition of American Manufactures* (Philadelphia: Franklin Institute, 1874), 230.

66. Francis A. Walker, ed., *United States Centennial Commission, International Exhibition, 1876: Reports and Award Groups XXI–XXVII*, vol. 7 (Washington, DC, 1880), 14–18; see also Coleman Sellers, "System of Awards at the United States International Exhibition of 1876," *JFI* 103 (1877): 14–16.

67. Fairmount Park Commission, Board of Commissioners Minutes, vol. 5 (October 14 and 23, 1876), CPA, Acc. 149.1; "Exhibition Echoes: A Conference as to What Shall Be Done with Memorial and Machinery Halls," *North American*, March 16, 1877, Fairmount Park Commission (FPC) Scrapbook, 1869–1903.

68. International Exhibition Company, "Official Bulletin of the International Exhibition" (Philadelphia, 1877), 4, HSP, Centennial Collection, Philadelphia Permanent Exhibition.

69. William Sellers & Co. Visitors Register, 1853–89, Hagley, Soda House.

70. *American Machinist* 1, nos. 5 and 12 (March and October 1878); vol. 2, no. 1 (November 1878); "Our American Mechanical Engineers. No. I. William Sellers," *American Machinist* 2, no. 8 (June 28, 1879): 7. See also J. P. Crittenden and Charles B. Helffrich, *Philadelphia Securities: A Descriptive and Statistical Manual of the Corporations of the City of Philadelphia* (Philadelphia: Burk & McFetridge, 1892), 716.

71. Typescript for *Railway Age*—American Railway Appliance Exposition (May 1905), William Sellers & Co. Scrapbook, Pennsylvania State Archives.

72. Outerbridge and Sellers, "William Sellers," 380.

73. I. L. Vansant, ed., *The Royal Road to Wealth, an Illustrated History of the Successful Business Houses of Philadelphia* (Philadelphia: Loag, c. 1868), 18.

74. Quoted in Charles Robson, *The Manufactories and Manufacturers of Pennsylvania of the Nineteenth Century* (Philadelphia: Galaxy, 1875), 16.

75. Ibid.

76. See, for example, Harvey Kantor and David B. Tyack, eds., *Work, Youth, and Schooling: Historical Perspectives on Vocationalism in American Education* (Stanford, CA: Stanford University Press, 1982); Lawrence A. Cremin, *American Education: The Metropolitan Experience, 1876–1980* (New York: Harper & Row, 1988). This interpretation has largely been echoed by labor and industrial historians, for example, Walter Licht, *Getting Work: Philadelphia, 1840–1950* (Cambridge, MA: Harvard University Press, 1992).

77. Joseph Wickham Roe, *English and American Tool Builders* (New Haven, CT: Yale University Press, 1916), 252; "Coleman Sellers Dies in Bryn Mawr," *Public Ledger*, August 16, 1922; "Howard Sellers: Member of the Union League and Franklin Institute to Be Buried Tomorrow," *Philadelphia Inquirer*, October 16, 1929.

78. Robert Kanigel, *The One Best Way: Frederick Winslow Taylor and the Enigma of Efficiency* (New York: Viking, 1997), 146.

79. Ibid., 233–34; Roe, *English and American Tool Builders*, 257–58.

80. Howell John Harris, *Bloodless Victories: The Rise and Fall of the Open Shop in the Philadelphia Metal Trades, 1890–1940* (New York: Cambridge University Press, 2000), 25, 43; Kanigel, *One Best Way*, 329.

81. *Millbourne Mills Company: Its Antecedents and Present Operations* (Philadelphia, 1888), 12–13; see also, Henry Graham Ashmead, *History of Delaware County, Pennsylvania* (Philadelphia: Everts, 1884), 547; S. F. Hotchkin, *Rural Pennsylvania: In the Vicinity of Philadelphia* (Philadelphia: Jacobs, 1897), 309.

82. Ellen Cronin, "The Sellers Family of Upper Darby: An Eighteenth and Nineteenth Century American Dynasty of Inventors, Millers and Mechanic-Manufacturers" (typescript in the possession of the author), 21–22.

83. *Transactions of the American Society of Mechanical Engineers*, vol. 8 (1887), v–xxv. See also Scranton, *Endless Novelty*, 104.

84. John K. Brown, "When Machines Became Gray and Drawings Black and White: William Sellers and the Rationalization of Mechanical Engineering," *IA: The Journal of the Society for Industrial Archeology* 25, no. 2 (1999): 43.

85. Franklin Institute, Board of Managers Minutes, vol. 2 (1848–70), 448, 460.

86. Outerbridge and Sellers, "William Sellers," 368.

87. Coleman Sellers, "The Worcester County Free Institute of Industrial Science," *JFI* 65, no. 4 (April 1873): 221.

88. Nina de Angeli Walls, "Art and Industry in Philadelphia: Origins of the Philadelphia School of Design for Women, 1848 to 1876," *PMHB* 127, no. 3 (July 1993): 177–99.

89. *Thirty-First Annual Report of the Board of Managers* (Philadelphia: Spangler & Davis, 1882); *Thirty-Fourth Annual Report of the Board of Managers* (Philadelphia: Spangler & Davis, 1885); *Thirty-Eighth Annual Report of the Board of Managers* (Philadelphia: Spangler & Davis, 1889); Dawson, *Lives of the Philadelphia Engineers*, 211–19; Andrew Dawson, "The Workshop and the Classroom: Philadelphia Engineering, the Decline of Apprenticeship, and the Rise of Industrial Training, 1878–1900," *History of Education Quarterly* 39, no. 2 (Summer 1999): 143–60. For discussion of the Textile School, see Philip Scranton, *Proprietary Capitalism: The Textile Manufacture at Philadelphia, 1800–1885* (New York: Cambridge University Press, 1983), 405–13.

90. Philip Scranton, *Figured Tapestry: Production, Markets and Power in Philadelphia Textiles, 1885–1941* (New York: Cambridge University Press, 2002), 26–27; Dawson, "Workshop and the Classroom."

91. "Annual Address of William Sellers," *Proceedings of the Engineers' Club of Philadelphia* 7 (May 1890): 312.

92. Dawson, "Workshop and the Classroom," 144, 156.

93. Sellers, *American Supremacy in Applied Mechanics*, 52.

94. Ibid., 51.

95. Sellers, "President's Address," 679.

96. See, for example, Coleman Sellers, "Mechanics: An Introductory Lecture Delivered Before the Franklin Institute, Friday, November 9th, 1883," Franklin Institute; "A Pennsylvanian: Coleman Sellers," *Evening Bulletin*, July 21, 1924, Sellers Library, Scrapbooks, vol. 15.

97. Coleman Sellers's address in University of Pennsylvania, "Proceedings at the Public Inauguration of the Building Erected for the Departments of Arts and Sciences, October 11, 1872," University of Pennsylvania Archives (hereafter UPA), Richards Papers.

98. Furness, *F.R.*, 15.

99. UPA, General Administration Papers, 1820–1930, box 15, 1869–73, Archives General—1870–72, New Building folder; *Minute Book of the Building Committee for the Erection of the New Building*, 1870–72, UPA; William Sellers to Thomas W. Richards, March 30, 1870, January 26, March 17 and 25, and October 28, 1871, and Charles J. Stillé to Thomas W. Richards, February 28, 1870, UPA, Thomas Richards Papers.

100. University of Pennsylvania, Trustees Minutes, vol. 11, March 5, 1878, 435–36, UPA.

101. Ibid., June 1, 1875, 299.

102. Ibid., vol. 12, April 28, 1884, 123.

103. J. Vaughan Merrick and William Sellers to the Board of Trustees of the University of Pennsylvania (December 28, 1893) and Contracts between the Trustees of the University of Pennsylvania and Edge Moor Iron Co. (September 12, 1892), UPA, Miscellaneous Contracts; Announcement of mortgage bond, UPA, General Administration Collection (1830–1920); Richard Wood, acting treasurer University of Pennsylvania, to Edge Moor Iron Co. [William H. Connell, treasurer], December 12, 1892, and Wilson Brothers & Co. to Richard Wood, March 29, 1893, UPA, Contracts box, Light and Heat folder #2. See also George E. Thomas and David B. Brownlee, *Building America's First University: An Historical and Architectural Guide to the University of Pennsylvania* (Philadelphia: University of Pennsylvania Press, 2000), 73.

104. On scientific industrial research, see John Kenly Smith Jr., "The Scientific Tradition in American Industrial Research," *Technology and Culture* 31, no. 1 (January 1990): 121–31.

105. George E. Thomas, *William L. Price: From Arts and Crafts to Modern Design* (New York: Princeton Architectural Press, 2000), 121; Thomas and Brownlee, *Building America's First University*, 213–14.

106. University of Pennsylvania, Trustees Minutes, vol. 14 (October 2, 1900), 41.

107. Eda Kranakis, *Constructing a Bridge: An Exploration of Engineering Culture, Design, and Research in Nineteenth-Century France and America* (Cambridge, MA: MIT Press, 1997), 292.

108. "Injectors," *American Machinist* 1, no. 6 (April 1878): 2–4.

109. University of Pennsylvania, Trustees Minutes, vol. 11 (March 1, 1881), facing 570.

110. Ibid., facing 576. See also Edmund James, *Education of Business Men—I and II* (New York: American Bankers' Association, 1892).

111. University of Pennsylvania, Trustees Minutes, vol. 13, October 5, 1896, 385.

112. On the context of *The Philadelphia Negro*, see Michael B. Katz and Thomas J. Sugrue, eds., *W. E. B. DuBois, Race, and the City: "The Philadelphia Negro" and Its Legacy* (Philadelphia: University of Pennsylvania Press, 1998), esp. 1–37 and 103–25.

113. *An Appeal to the Citizens of Pennsylvania for the Foundation of a Veterinary Department in the University of Pennsylvania* (Philadelphia, 1879), 1–2, UPA.

114. Ibid., 5.

115. Ibid., 6.

5. Empires of Steel

1. Matthew Frye Jacobson, *Barbarian Virtues: The United States Encounters Foreign Peoples at Home and Abroad, 1876–1917* (New York: Hill & Wang, 2000), 16.

2. Robert Kanigel, *The One Best Way: Frederick Winslow Taylor and the Enigma of Efficiency* (New York: Viking, 1997), 161; see also *Pennsylvania and the Centennial Exposition* (Philadelphia: Gillin & Nagle, 1878), 113.

3. Coleman Sellers, "The Progress of the Mechanical Arts in Three Quarters of a Century," *JFI* 149 (January 1900): 5–25 (quotes from 21).

4. Ernest N. Paolino, *The Foundations of the American Empire: William Henry Seward and U.S. Foreign Policy* (Ithaca, NY: Cornell University Press, 1973), 39–40. For general discussions of the integration of the global economy in the late nineteenth century, see Kevin O'Rourke and Jeffrey Williamson, *Globalization and History: The Evolution of a Nineteenth-Century Atlantic Economy* (Cambridge, MA: MIT Press, 1999); Michael Bordo, Alan M. Taylor, and Jeffrey Williamson, *Globalization in Historical Perspective* (Chicago: University of Chicago Press, 2003).

5. Contemporary profiles of William Sellers's post–Civil War business development are found in "Our American Mechanical Engineers," *Manufacturer* 4, no. 1 (January 1891): 10.

6. For discussion of the range of late nineteenth- and early twentieth-century modes and strategies of industrial production and business management, see Philip Scranton, *Proprietary Capitalism: The Textile Manufacture at Philadelphia, 1800–1885* (New York: Cambridge University Press, 1983); Philip Scranton, *Endless Novelty: Specialty Production and American Industrialization, 1865–1925* (Princeton, NJ: Princeton University Press, 1997); Charles Sabel and Jonathan Zeitlin, eds., *World of Possibilities: Flexibility and Mass Production in Western Industrialization* (New York: Cambridge University Press, 1997); John K. Brown, *The Baldwin Locomotive Works, 1831–1915: A Study in American Industrial Practice* (Baltimore: Johns Hopkins University Press, 1995). These authors all write against the limits of the portrayal of industrial production in Alfred D. Chandler Jr., *The Visible Hand: The Managerial Revolution in American Business* (Cambridge, MA: Harvard University Press, 1977).

7. Geoffrey Tweedale, *Sheffield Steel and America: A Century of Commercial and Technological Interdependence, 1830–1930* (New York: Cambridge University Press, 1987), 114.

8. Coleman Sellers to George Escol Sellers, October 21, 1865, APS, Sellers Family Papers.

9. John K. Brown, correspondence with the author, May 15, 2004.

10. *JFI* 80 (November 1865): 357.

11. Ibid.

12. Jeanne McHugh, *Alexander Holley and the Makers of Steel* (Baltimore: Johns Hopkins University Press, 1980), 211; also Thomas Misa, *A Nation of Steel: The Making of Modern America, 1865–1925* (Baltimore: Johns Hopkins University Press, 1995), esp. 21–22.

13. Charles Wrege and Ronald G. Greenwood, "Origins of Midvale Steel (1866–1880): Birthplace of Scientific Management," in *Essays in Economic and Business History*, vol. 7, ed. Edwin J. Perkins (Economic and Business Historical Society, 1989), 205–19 (quote is from 207).

14. Franklin Institute, Board of Managers Minutes, vol. 2, April 10, 1867, 388; Tweedale, *Sheffield Steel and America*, 115.

15. R. G. Dun & Co., credit report on W. Butcher & Co. (September 18, 1867), PA vol. 11, p. 67.

16. Midvale Steel Co., *Fiftieth Anniversary, 1867–1917* (Philadelphia, c. 1917), 12.

17. R. G. Dun & Co., credit report on the William Butcher Steel Works (June 7, 1869), PA vol. 11, p. 67.

18. John A. Kouwenhoven, "The Designing of the Eads Bridge," *Technology and Culture* 23, no. 4 (October 1982): 535–68 (quote is on 560). Thanks also to John K. Brown for sharing his understanding of the roles of Butcher and the Philadelphia Interests in this project.

19. Andrew Carnegie to James Eads, October 18, 1870, quoted in Kouwenhoven, "Designing of the Eads Bridge," 560.

20. Quoted in Tweedale, *Sheffield Steel and America*, 115.

21. Ibid., 115.

22. John K. Brown, correspondence, May 15, 2004.

23. R. G. Dun & Co., credit reports on the William Butcher Steel Works (July 15, August 30, and November 7, 1871), PA vol. 11, pp. 67, 448.

24. Kouwenhoven, "Designing of the Eads Bridge," 560–63; Tweedale, *Sheffield Steel and America*, 115–16 (quote is again from Aertsen).

25. Wrege and Greenwood, "Origins of Midvale Steel," 209.

26. R. G. Dun & Co., credit report on Midvale Steel Works, November 30, 1871, PA vol. 11, p. 448.

27. Wrege and Greenwood, "Origins of Midvale Steel," 209.

28. R. G. Dun & Co., credit report on Midvale Steel Works, May 25, 1873, and July 29, 1874, PA vol. 11, p. 448.

29. "Samuel Huston vs. E. W. Clark, et al.," Israel H. Johnson's Minority Report (1876), p. 6, Reading Railroad file, HSP; Kanigel, *One Best Way*, 158.

30. R. G. Dun & Co., credit report on Midvale Steel Works, August 18, 1876, PA vol. 11, p. 449.
31. Wrege and Greenwood, "Origins of Midvale Steel."
32. R. G. Dun & Co., credit report on Midvale Steel Works, March 21, 1878, and September 19, 1879, PA vol. 11, p. 449.
33. Ibid., June 23 and October 27, 1880.
34. Tweedale, *Sheffield Steel and America*, 116; see also an account of this work based upon interviews with Brinley and William Sellers, in Henry Roland, "Six Examples of Successful Shop Management," *Engineering Magazine* 12, no. 5 (February 1897): 831–37.
35. Charles A. Brinley, *Russell Wheeler Davenport: Father of Rowing at Yale; Maker of Guns and Armor Plate Used by the United States Government* (New York: Putnam's Sons, 1905), 35.
36. Ibid., 36–38.
37. Ibid.
38. Ibid., 38.
39. Kanigel, *One Best Way*, 158.
40. Brinley, *Russell Wheeler Davenport*, 39–40.
41. Ibid.
42. Ibid.
43. Midvale Steel, journal of activities and memos from the Superintendent (1876–87), 9, Franklin Institute, Midvale Steel Collection.
44. Ibid.
45. Daily Notes of Charles A. Brinley (1873–82), Franklin Institute, Midvale Collection.
46. Ibid.
47. Kanigel, *One Best Way*, 184.
48. For general works on Taylor and Taylorism, see Daniel Nelson, *Frederick W. Taylor and the Rise of Scientific Management* (Madison: University of Wisconsin Press, 1980); Kanigel, *One Best Way*.
49. For discussion of scientific management's adoption in both manufacturing and other businesses, see Daniel Nelson, ed., *A Mental Revolution: Scientific Management since Taylor* (Columbus: Ohio State University Press, 1992).
50. Andrew Dawson argues likewise: "Taylor claimed scientific management as the product of his own genius but a more fruitful approach is to situate Taylor within the contemporary industrial milieu—to see him as much an outcome of circumstances as an initiator of change." Andrew Dawson, *Lives of the Philadelphia Engineers: Capital, Class and Revolution, 1830–1890* (Burlington, VT: Ashgate, 2004), chap. 7, p. 3.
51. Brinley, *Russell Wheeler Davenport*, 45.
52. Wrege and Greenwood, "Origins of Midvale Steel," 211.
53. Misa, *Nation of Steel*, 181–82; Kanigel, *One Best Way*, 176–79.
54. Tweedale, *Sheffield Steel and America*, 68.
55. Roland, "Six Examples of Successful Shop Management," 835.
56. Ibid., 836.
57. Ibid. (June 21, 1879).
58. Ibid.
59. Ibid.
60. McHugh, *Alexander Holley and the Makers of Steel*, 308.
61. Misa, *Nation of Steel*, 36; "Discussion on Steel Rails," *American Institute of Mining Engineers Transactions* 9 (1880): 529–608; *Engineering and Mining Journal* 25 (June 8, 1878): 396.
62. Misa, *Nation of Steel*, 36.
63. William Sellers to Joseph Wharton, c. 1892, Swarthmore College Friends Historical Library, Wharton Papers.
64. Richard D. Glasgow, *Prelude to a Naval Renaissance: Ordnance Innovation in the United States during the 1870s* (dissertation, University of Delaware, 1878).

65. Thomas R. Heinrich, *Ships for the Seven Seas: Philadelphia Shipbuilding in the Age of Industrial Capitalism* (Baltimore: Johns Hopkins University Press, 1997), 99.

66. Estimates for Shafting and Machine Tools, 1862–72, William Sellers & Co. Records, Item 3, Hagley, Soda House.

67. Frankford Arsenal Records, register of letters, telegrams, & endorsements received (1864–1903), National Archives, Philadelphia, Record Group 156.

68. Brinley, *Russell Wheeler Davenport*, 40.

69. Midvale Steel Co. to Chief of Ordnance, July 22, August 22 and 23, September 5, 17, 21, and October 7, 8, 29, 31, 1881, Frankford Arsenal Records, register of letters, telegrams, & endorsements received, National Archives, Philadelphia, Record Group 156. Also, Memoranda re—carriage castings, plates & c. (1882–86), pp. 9, 13, 15, 21, 25, 41, 57, 70; Record of High Power Steel Breech-Loading Rifles (1882–90), pp. 6, 38–46.; Gun Forgings—Foreign and Domestic (1883–87), pp. 15, 31, 35–38, 45, 49, 53, 101–2, 113, 117, all National Archives, Washington, DC, Record Group 74.

70. John K. Brown, correspondence, May 15, 2004.

71. See, for example, Rough Record of Tests of Steel &c. with Reference to the Woodbridge 8 in. Breech-Loading Rifles, vol. 3 (1883–84), pp. 5, 11, 13, 16–20, National Archives, Washington, DC, Record Group 74.

72. Howard C. Myers Jr., Notes on Midvale History (1957), Franklin Institute, Midvale Steel Collection.

73. Brinley, *Russell Wheeler Davenport*, 48.

74. R. W. Davenport to Coleman Sellers, December 22, 1883, APS, Sellers Family Papers.

75. Coleman Sellers to Horace, Cora, and Coleman Sellers Jr., June 20, 1884, Coleman Sellers letter books, APS, Sellers Family Papers.

76. Coleman Sellers to Coleman Sellers Jr., June 11, 1884, Coleman Sellers letter books, APS, Sellers Family Papers.

77. William Sellers to Coleman Sellers, July 24, 1884, Coleman Sellers letter books, APS, Sellers Family Papers.

78. Gun Foundry Board, Commandant's Office, U.S. Navy Yard, League Island, PA, to W. E. Chandler, secretary of the navy, May 19, 1883; Gun Foundry Board, Memorandum of Recommendations Contained in Supplementary Report, 1883, both Hagley, Soda House, B. Franklin Cooling Research Notes.

79. Record of High Power Steel Breech-Loading Rifles (1882–90); Memoranda re—carriage castings, plates & c. (1882–86); Gun Forgings—Foreign and Domestic (1883–87).

80. Gun Forgings—Foreign and Domestic (1883–87), p. 153.

81. For the end of the nineteenth century, see Record of Manufacture of 6 in. Breech-Loading Rifles at Midvale Steel (1887–91); Index to Gun Forgings at Midvale (1896–97); Record of Hotchkiss Guns (1887–91); Record of Carriage Castings (1889–93), pp. 260–61, 284–91, all National Archives, Washington, DC, Record Group 74.

82. William Sellers & Co. to Chief of Ordnance, September 6 and November 15, 1877; April 19 and 22, September 19, October 15, 1878; October 2 and 8, 1879; February 11 and December 31, 1880; January 13, 14, 19, 20, April 26, 28, May 2, 1881, Records of the Office of the Chief of Ordnance, Frankford Arsenal Records, register of letters, telegrams, & endorsements received, National Archives, Philadelphia, Record Group 156.

83. William Sellers, "Machinery Manufacturing Interests," in *One Hundred Years of American Commerce*, vol. 2, ed. Chauncey M. Depew (New York: Haynes, 1895), 349.

84. See, for example, Wm. Sellers & Co. to Chief of Ordnance, April 25, 1881, Records of the Office of the Chief of Ordnance, Frankford Arsenal Records, register of letters, telegrams, & endorsements received.

85. Alexander E. Outerbridge Jr. and Coleman Sellers Jr., "William Sellers," *JFI* 159 (May 1905): 372. This same incident is also recounted in Joseph Wickham Roe, *English and American Tool Builders* (New Haven, CT: Yale University Press, 1916), 250.

86. Philip Scranton, "Large Firms and Industrial Restructuring: The Philadelphia Region 1900–1980," *Pennsylvania Magazine of History and Biography* 116 (1992): table 1.

87. See, for example, Franklin Institute, Board of Managers Minutes, January 10, 1883.

88. Myers, Notes on Midvale History; Kanigel, *One Best Way*, 236, 240; J. P. Crittenden and Charles B. Helffrich, *Philadelphia Securities: A Descriptive and Statistical Manual of the Corporations of the City of Philadelphia* (Philadelphia: Burk & McFetridge, 1892), 100.

89. Taylor would carry forward his work in scientific management, initially among Philadelphia metalworkers, then at Bethlehem Steel, and finally in a consulting practice that disseminated Taylorism in some two hundred companies between 1901 and his death in 1915. In the mid-1890s, he continued his metal-cutting experiments as a consultant for Cramp's shipyard, I. P. Morris, and Sellers & Co., all of which were busy with big military jobs. Cramp and Sellers shared the cost of tests, most of them conducted at the Bush Hill plant, to determine which type of steel made the best cutting tools and what speed was optimal for running Sellers-made lathes, primarily for ship and engine production. At Sellers & Co., in a familiar project, Taylor helped refine piece-rates through tests on tool steel. But his time at the tool works points up some of scientific management's limits. Experienced workers there resisted his brief and overzealous attempt at wholesale reorganization of workshop management, leading William to terminate his consultancy. Midvale's huge orders made for repetitive work and gave incentives to control the minutia of shop floor labor down to the smallest movements of workers' arms and legs. For Sellers & Co., over-rationalizing production had serious costs. Craft knowledge and discretion remained important for its specialized jobs and its capacity to innovate. J. Sellers Bancroft to F. W. Taylor, May 25, 1894, and F. W. Taylor to J. Sellers Bancroft, May 25, 1895, both in Gantt-Taylor Correspondence, Hagley, Soda House (copies of originals in the Taylor Collection at the Stevens Institute of Technology); Kanigel, *One Best Way*, 273; Misa, *Nation of Steel*, 182–83; Dawson, *Lives of the Philadelphia Engineers*, chap. 7, pp. 18–21; John K. Brown, "When Machines Became Gray and Drawings Black and White: William Sellers and the Rationalization of Mechanical Engineering," *IA: The Journal of the Society for Industrial Archeology* 25, no. 2 (1999): 46.

90. The central features that Taylor ascribed to his system were all in evidence at Sellers & Co. and Midvale before his arrival in Nicetown, though he certainly did important research there and thereafter. Nelson, *Frederick Winslow Taylor*, 102.

6. Building the Scientific City

1. Coleman Sellers, "Scientific Method in Mechanical Engineering," *JFI* 120 (December 1885): 420–38 (quotes from 436).

2. For detailed accounts of the nineteenth-century urban reform movements that constitute the prehistory of city planning in the United States, see, for example, Paul Boyer, *Urban Masses and Moral Order in America, 1820–1920* (Cambridge, MA: Harvard University Press, 1978); Jon Peterson, *The Birth of City Planning in the United States, 1840–1917* (Baltimore: Johns Hopkins University Press, 2003).

3. Joseph Wickham Roe, *English and American Tool Builders* (New Haven, CT: Yale University Press, 1916), 248.

4. Quoted in Alexander E. Outerbridge Jr. and Coleman Sellers Jr., "William Sellers," *JFI* 159 (May 1905): 370–71.

5. *Report of the Twenty-Seventh Exhibition of American Manufactures* (Philadelphia: Franklin Institute, 1874), 231.

6. Coleman Sellers Jr., "Mechanical Drawing: A lecture Delivered before the Franklin Institute, November 11th, 1881," *JFI* 113 (January 1882), 32–49.

7. Domenic Vitiello, "Engineering the Metropolis: William Sellers, the Wilson Bros., and Industrial Philadelphia," *Pennsylvania Magazine of History and Biography* 126, no. 2 (2002): 273–303; George E. Thomas, "'The Happy Employment of Means to Ends': Frank Furness's Library of the University of Pennsylvania and the Industrial Culture of Philadelphia," *Pennsylvania Magazine of History and Biography* 126, no. 2 (2002).

8. William Sellers & Co. [Justus C. Schwacke] to John S. Albert, Esq., Chief of the Bureau of Machinery, December 10, 1875; William Sellers & Co. [H.C. Francis] to John S. Albert, Esq., January 7, 1876; William Sellers & Co. [H.C. Francis], "Proposal for Shafting" (n.d.), all in City of Philadelphia Archives (hereafter CPA), Record Group 230.29, Centennial Commission, Bureau of Machinery, Construction Papers, Proposals folder; see also *American Machinist* 1, no. 5 (March 1878): 1–3.

9. John McArthur Jr. and Joseph M. Wilson, "Revised Design in Competition for Proposed Centennial Exhibition Building, for Exposition of Industry of All Nations" (Philadelphia, 1873), 17, and Henry Pettit and Joseph M. Wilson, "Specifications for the Construction, Erection, and Finishing of the Main Exhibition Building on Lansdowne Plateau, Fairmount Park" (Philadelphia, 1874), both in HSP, *Documentary Record of the Centennial Exhibition*, vol. 3b, Buildings.

10. "Catalogue for the Sale of the Building of the International Exhibition Company," March 28, 1881, 7, HSP, Centennial Collection, Philadelphia Permanent Exhibition; "Under the Hammer: Selling Centennial Relics," *Philadelphia Evening Bulletin*, October 13, 1881, 1; Ernest C. Miller, ed., *This Was Early Oil: Contemporary Accounts of the Growing Petroleum Industry, 1848–1885* (Harrisburg, PA, 1968), 53.

11. The firm was incorporated with a substantial initial capital of $1 million. *An Act to Incorporate the Edge Moor Iron Company* (January 20, 1869), Hagley, Soda House, Edge Moor Iron Co. material. See also R.G. Dun, Edge Moor Iron Works credit reports, Delaware Reel, vol. 1, p. 277.

12. William Sellers & Co., *A Treatise on Machine-Tools* (Philadelphia: J.B. Lippincott, 1877), 263.

13. Edge Moor Iron Co., "The Galloway Boiler" (January 1885); Edge Moor Iron Co., *Catalogue of the Edge Moor Iron Co.*, 1902, 5.

14. The quote is from Outerbridge and Sellers, "William Sellers," 366. See also "Plate Shearing Machine," *American Machinist* 1, no. 2 (December 1877): 1; Washington Roebling to John Roebling, October 11, 1868, Rensselaer Polytechnic Institute (RPI), Roebling Papers (this last source comes by way of John K. Brown).

15. William Sellers to Coleman Sellers, July 24, 1884, Coleman Sellers letter books, APS, Sellers Family Papers.

16. Thomas Scharf, *History of Delaware, 1609–1888* (Philadelphia: Richards, 1888), 805.

17. Joseph M. Brumbley Sr., *Where the Pigeons Slept: A History* (Wilmington, DE, 1989), 136.

18. Roe, *English and American Tool Builders*, 250. For an overview of European and American bridge-building history, see Joseph Gies, *Bridges and Men* (New York: Grosset & Dunlap, 1963); see also Tom F. Peters, *Building the Nineteenth Century* (Cambridge, MA: MIT Press, 1996).

19. Coleman Sellers, *American Supremacy in Applied Mechanics: A Series of Papers Reprinted from the Engineering Magazine* (New York: Engineering Magazine Co., 1892), 35, APS, Sellers Family Papers.

20. Ibid., 35–36.

21. Anna T. Lincoln, *Wilmington, Delaware, Three Centuries under Four Flags, 1609–1937* (Port Washington, NY: Kennikat, 1972); "Wilmington and its Industries," *Lippincott's Magazine*, April 1873, 377–78.

22. Brumbley, *Where the Pigeons Slept*, 135–37.

23. Various deeds to Eli Garrett, December 30, 1868, and from Eli Garrett to Edge Moor Iron Co., June 17, 1869, both Hagley, Soda House, Edge Moor Iron Co. materials. See also deed from Philadelphia, Wilmington & Baltimore Railroad to William Sellers,

November 1, 1867; Samuel Smedley to I. C. Price, November 10, 1868; and agreement between Philadelphia, Wilmington & Baltimore Railroad and William Sellers, May 1, 1869, all HSD, Sellers-Cox Papers.

24. "Death of George H. Sellers" (June 9, 1897), Sellers Library, Sellers Scrapbooks, vol. 11, p. 141.

25. Mortgage: The Edge Moor Iron Co. to the Pennsylvania Company for Insurance on Lives and Granting Annuities, July 1, 1875, Hagley, Soda House, Edge Moor Iron Co. material.

26. *The Wilson Brothers & Co.: A Catalogue of Works Executed* (Philadelphia, 1885); *American Machinist* 1, no. 8 (June 1878): 10, and vol. 2, no. 20 (September 1879): 10; "Our American Mechanical Engineers. No. I. William Sellers," *American Machinist* 2, no. 8 (June 28, 1879): 7; A. J. Clement, *Wilmington, Delaware: Its Productive Industries and Commercial and Maritime Advantages* (Wilmington: Delaware Printing Co., 1888), 31–32; *Delaware's Industries: An Historical and Industrial Review, 1891* (Philadelphia: Keighton, 1891); W. W. Evans to Coleman Sellers, July 19, 1886, APS, Sellers Family Papers; Scharf, *History of Delaware*, 805.

27. R. G. Dun, Edge Moor Bridge Works credit report, April 3, 1889, Delaware Reel, vol. 2, p. 400.

28. P. A. C. Spero & Co., *Delaware Historic Bridges Survey and Evaluation* (DelDOT, 1991); Historic American Engineering Record, DE-48, PA-277-C, PA-544, RI-52.

29. Scharf, *History of Delaware*, 805.

30. Edge Moor Iron Co. catalog, 1902, 118–20, Hagley Library.

31. Quoted in Brumbley, *Where the Pigeons Slept*, 131.

32. R. G. Dun, Edge Moor Iron Co. credit report, April 3, 1889, Delaware Reel, vol. 2, p. 400.

33. Thomas Misa, *A Nation of Steel: The Making of Modern America, 1865–1925* (Baltimore: Johns Hopkins University Press, 1995); Eli Imberman, "The Formative Years of the Chicago Bridge and Iron Company" (PhD dissertation, University of Chicago, 1973); Victor C. Darnell, *Directory of American Bridge Building Companies, 1840–1900* (Society for Industrial Archaeology, 1984).

34. "Large Iron Contract and Its Terms," *PM* (July 12, 1879), RPI, Roebling Papers; see also Charles Wrege and Ronald G. Greenwood, "Origins of Midvale Steel (1866–1880): Birthplace of Scientific Management," in *Essays in Economic and Business History*, vol. 7, ed. Edwin J. Perkins (Economic and Business Historical Society, 1989), 210–11; David McCullough, *The Great Bridge: The Epic Story of the Building of the Brooklyn Bridge* (New York: Simon & Schuster, 1972), 372, 382–96, 409–10, 434–51; Brumbley, *Where the Pigeons Slept*, 159–64; "The Bridge: The Conference Committee and the Steel Contract," *Brooklyn Daily Eagle*, June 27, 1882; "A Change in the Bridge Management Imperatively Required," *Brooklyn Daily Eagle*, July 11, 1882; "Roosevelt and the Bridge," *New York Herald*, June 16, 1882; "The Bridge Investigation," *New York Herald*, June 17, 1882; "The Bridge: An Important Secret Conference This Morning," *Brooklyn Daily Eagle*, June 21, 1882; "Completing the Span," *New York Herald*, June 22, 1882; "Colonel Sellers and the Bridge," *Brooklyn Union and Argus*, June 22, 1882.

35. John K. Brown, correspondence with the author, May 15, 2004.

36. Franklin Institute, Board of Managers Minutes, January 12, 1887.

37. The classic study of the parks movement and its impacts on urban and suburban development is David Schuyler, *The New Urban Landscape: The Redefinition of City Form in Nineteenth-Century America* (Baltimore: Johns Hopkins University Press, 1986). See also Peterson, *Birth of City Planning in the United States*.

38. Thomas Daniels defines environmental planning as "the theory and practice of making good, interrelated decisions about the natural environment (natural resources, wildlife, and natural hazards), working landscapes (farms, forests, and lands from which minerals

are extracted), public health (air and water pollution, toxics, and waste disposal), and the built environment." Thomas L. Daniels, "A Trail across Time: American Environmental Planning from City Beautiful to Sustainability," *Journal of the American Planning Association* 75, no. 2 (Spring 2009): 178–92, quote on 178.

39. Record of Select Council, October 26, 1854, CPA.

40. Fairmount Park Commission, Board of Commissioners Minutes, vol. 1 (1867–69), CPA, Acc. 149.1; *First Annual Report of the Commissioners of Fairmount Park* (Philadelphia: King & Baird, 1869).

41. Fairmount Park Commission, Board of Commissioners Minutes, vol. 1, October 11, 1867, CPA, Acc. 149.1; see also Henry P. M. Birkinbine, "The Future Water Supply of Philadelphia," *JFI* 105, no. 5 (1878).

42. Fairmount Park Commission, Board of Commissioners Minutes, vol. 1, October 11, 1867, CPA, Acc. 149.1; see also Birkinbine, "The Future Water Supply of Philadelphia."

43. Fairmount Park Commission, Board of Commissioners Minutes, vol. 1, December 31, 1868, CPA.

44. Ibid.

45. Ibid., vol. 2, June 5, 1869.

46. Ibid., vols. 3–4, 1871–76; Committee on Plans and Improvements, Minutes, vol. 2 (1872–79), CPA, Acc. 149.17.

47. Fairmount Park Commission, Board of Commissioners Minutes, vol. 1, December 21, 1867, CPA.

48. Ibid., vol. 3, November 11, 1871.

49. *Third Annual Report of the Commissioners of Fairmount Park* (Philadelphia: King & Baird, 1871), 28–29, CPA.

50. Fairmount Park Commission, Board of Commissioners Minutes, vol. 1, July 6 and 26, 1867, and vol. 3, December 14, 1872, CPA.

51. For an account of the class dimensions of the late nineteenth-century parks movement, see Elizabeth Blackmar and Roy Rosenzweig, *The Park and the People: A History of Central Park* (Ithaca, NY: Cornell University Press, 1992). See also Frederick Law Olmsted, *Civilizing American Cities: Writings on City Landscapes*, ed. S. B. Sutton (Cambridge, MA: MIT Press, 1971).

52. Fairmount Park Art Association, *Seventh Annual Report of the Board of Trustees and List of Members* (Chandler, 1879), 16; see also Fairmount Park Art Association, *Charter, By-Laws, and Lists of Officers and Members* (J. B. Lippincott, 1872), both David W. Sellers Scrapbooks, vol. 2, Fairmount Park Commission Archives.

53. Fairmount Park Commission, Board of Commissioners Minutes, vol. 1, July 13 and September 7, 1867, and vol. 2, April 9, 1870, CPA.

54. Eli K. Price letter to the editor of the *Philadelphia Press*, May 26, 1868, in Committee on Land Purchases, Minutes, 1867–83, CPA, Acc. 149.16.

55. Eli K. Price, "Report on the Acquisition of Lands for Fairmount Park," April 30, 1877, in Committee on Land Purchases, Minutes, 1867–83, CPA, Acc. 149.16.

56. Fairmount Park Commission, Board of Commissioners Minutes, vol. 2, CPA.

57. *Fourth Annual Report of the Commissioners of Fairmount Park*, 13, CPA.

58. In the Court of Common Pleas of Philadelphia County, *Sellers et al. vs. the Pennsylvania Railroad Company et al.* (December Term, 1874), 1–2, Library Company of Philadelphia (LCP).

59. Ibid., 34–35.

60. *The Philadelphia Stock Yards and Abattoir: The Testimony in Favor of their Location on the Schuylkill River above Market Street* (1875), 4, LCP.

61. Ibid., 9–10.

62. Ibid., 10.

63. Catherine Brinkley and Domenic Vitiello, "From Farm to Nuisance: Animal Agriculture and the Rise of City Planning Regulation," *Journal of Planning History* (forthcoming).

64. Franklin Institute, Board of Managers Minutes, December 13, 1882, FI.

65. Minutes of the Commissioners for the Erection of the Public Buildings, III (1895), CPA, Record Group 160.1; Expenditures Journal (1872–1901), CPA, Record Group 160.12; City Hall Drawings and Plans (1871–1901), CPA, Record Group 160.20.

66. Summary of Joseph Wilson's discussion of Edward Clapp Shankland, "Steel Skeleton Construction in Chicago," *Minutes of Proceedings of the Institution of Civil Engineers* 128 (London, 1897): 54–55; Vitiello, "Engineering the Metropolis." Architectural histories that locate the first skyscrapers in Chicago include Sigfried Giedion, *Space, Time and Architecture: The Growth of a New Tradition* (Cambridge, MA: Harvard University Press, 1949); Henry Russell Hitchcock and Philip Johnson, *The International Style: Architecture since 1922* (New York: W. W. Norton, 1932). See also George E. Thomas, *William L. Price: From Arts and Crafts to Modern Architecture* (New York: Princeton Architectural Press, 2000).

67. Misa, *Nation of Steel*, 47–48, 300 n. 9, 64–74.

68. Sellers, *American Supremacy in Applied Mechanics*, 34–35; see also Barr Fevree to Coleman Sellers, July 10, 1895, APS, Sellers Family Papers.

69. *Gopsill's Philadelphia City Directory for 1890* (Philadelphia: Gopsill, 1890).

70. For an account of the variety and evolution of patterns of industrial district formation in another North American city during the late nineteenth and early twentieth centuries, see Robert Lewis, *Manufacturing Montreal: The Making of an Industrial Landscape, 1850 to 1930* (Baltimore: Johns Hopkins University Press, 2000).

71. William Sellers & Co. to U.S. Mint, November 11 and 13, 1896, National Archives—Philadelphia, RG 104, U.S. Mint, Specifications for Machinery and General Supplies.

72. John K. Brown, *The Baldwin Locomotive Works, 1831–1915: A Study in American Industrial Practice* (Baltimore: Johns Hopkins University Press, 1995), 191. Also, J. P. Crittenden and Charles B. Helffrich, *Philadelphia Securities: A Descriptive and Statistical Manual of the Corporations of the City of Philadelphia* (Philadelphia: Burk & McFetridge, 1892), 716.

73. "A Thoroughly Modern Injector Plant," *Locomotive Fireman's Magazine,* December 1906, in Scrapbook, PA State Archives, MG 315.

74. Brown, *Baldwin Locomotive Works*, 204; *Builders Guide* 17, no. 46 (November 12, 1902).

75. "Thoroughly Modern Injector Plant."

76. By World War I, factories in the inner city consisted largely of multistory steel and concrete structures up to twelve or more stories in height. This building type proliferated further in the 1920s. Suburban plants would predominate after World War II, accelerating the suburbanization of industry and causing later historians to largely ignore the early twentieth-century vertical push of factories in the city. One exception is Amy Slaton, *Reinforced Concrete and the Modernization of American Building, 1900–1930* (Baltimore: Johns Hopkins University Press, 2001).

77. The Philadelphia Social History Project remains the classic quantitative study of the urban ecology of the late nineteenth century. Theodore Hershberg, ed., *Philadelphia: Work, Space, Family, and Group Experience in the 19th Century* (New York: Oxford University Press, 1981); Stephanie W. Greenberg, "The Relationship between Work and Residence in an Industrializing City: Philadelphia, 1880," in *The Divided Metropolis: Social and Spatial Dimensions of Philadelphia, 1800–1975*, ed. William W. Cutler III and Howard Gillette Jr. (Westport, CT: Praeger, 1980), 141–68.

78. *McElroy's Philadelphia City Directory for 1860* (Philadelphia: Biddle, 1860).

79. Deed Book LRB 188, p. 442, June 14, 1866, CPA.

80. *Gopsill's Philadelphia City Directory for 1870* (Philadelphia: Gopsill, 1870); *Gopsill's Philadelphia City Directory for 1880* (Philadelphia: Gopsill, 1880); *Gopsill's Philadelphia City Directory for 1890*.

81. Deed abstracts, plots 4 N 20–4–6, 12, 48, 52, 60, CPA. *Gopsill's Philadelphia City Directory for 1900*, 178, 2058, 2333.

82. For an account of Chestnut Hill, see David Contosta, *Suburb in the City: Chestnut Hill, Philadelphia, 1850–1990* (Columbus: Ohio State University Press, 1992).

83. For discussion of late nineteenth-century "garden suburbs," see Mary Corbin Sies, "The City Transformed: Nature, Technology, and the Suburban Ideal, 1877–1917," *Journal of Urban History* 14, no. 1 (1987): 81–111; Mary Corbin Sies, "North American Suburbs, 1880–1950: Cultural and Social Reconsiderations," *Journal of Urban History* 27, no. 3 (2001): 313–46; Schuyler, *New Urban Landscape*; John R. Stilgoe, *Borderland: Origins of the American Suburb, 1820–1939* (New Haven, CT: Yale University Press, 1988).

84. "Through William Penn's 'Low Counties,'" *Lippincott's Magazine of Popular Literature and Science,* September 1872, 258.

85. Ibid., 256. See also J. L. Smith, *Atlas of Properties along the Philadelphia, Wilmington and Baltimore R.R., Baltimore & Ohio R.R., Phila. & West-Chester R.R., and part of Philada. & Reading R.R., Philadelphia to Chester and Elwyn Sta.* (Philadelphia: Smith, 1889), plates 20 and 23.

86. Quoted in "Through William Penn's 'Low Counties,'" 256. "Interview with Frank Kane," *Chester Times,* 1925, Ridley Park Historical Society.

87. "Through William Penn's 'Low Counties,'" 256.

88. Ibid., quoted p. 256.

89. For a review of these patterns' endurance, see Mary Corbin Sies, "Paradise Retained: An Analysis of Persistence in Planned, Exclusive Suburbs, 1880–1980," *Planning Perspectives* 12, no. 2 (1997): 165–91.

90. Disston's development of the Tacony neighborhood is recounted in detail in Harry C. Silcox, *A Place to Live and Work: The Henry Disston Saw Works and the Tacony Community of Philadelphia* (University Park, PA: Pennsylvania State University Press, 1994). Dobson and Powers & Weightman's investments in company housing in their East Falls neighborhood have not been well documented by historians, though evidence from fire insurance atlases and deeds reveals the general patterns. See, for example, Elvino Smith, *Atlas of the City of Philadelphia* (Philadelphia, 1875).

91. This argument is elaborated in Robert Lewis, ed., *Manufacturing Suburbs: Building Work and Home on the Metropolitan Fringe* (Philadelphia: Temple University Press, 2004).

92. See, for example, Daniel T. Rodgers, *Atlantic Crossings: Social Politics in a Progressive Age* (Cambridge, MA: Harvard University Press, 1998).

93. Coleman Sellers to Coleman Sellers Jr., June 11, 1884, Coleman Sellers letter books, APS, Sellers Family Papers.

94. Silcox, *Place to Live and Work*; Margaret Crawford, *Building the Workingman's Paradise: The Design of American Company Towns* (New York: Verso, 1995); John Garner, ed., *The Company Town: Architecture and Society in the Early Industrial Age* (New York: Oxford University Press, 1992).

95. Sellers, *American Supremacy in Applied Mechanics*, 36. See also, Coleman Sellers to Coleman Sellers Jr., June 11, 1884.

96. Brumbley, *Where the Pigeons Slept*, 117–18.

97. Lincoln, *Wilmington, Delaware*, 363.

98. W. H. Wilson to William Sellers, March 1, 1882; deed from David H. Beeson to William Sellers, June 2, 1884; deed from William and Amelie Sellers to Philadelphia, Wilmington & Baltimore Railroad (n.d.); William Sellers to William F. Sellers, May 12, 15, 16, and 29 1888; William Sellers to William H. Brown, May 15, July 3, and September 3, 1888; William Sellers to William F. Sellers, May 15, 1888; agreement between William Sellers and Philadelphia, Wilmington & Baltimore Railroad Co., April 11, 1900, all HSD, Sellers-Cox Papers.

99. *An Act to Incorporate the Cherry Island Marsh Company of New Castle County* (Wilmington, DE: Roop, 1888), Hagley, Soda House, Edge Moor Iron Co. material.

100. Scharf, *History of Delaware*, 805.

101. Ibid.

102. Edge Moor Iron Athletic Association, annual outing programs, 1924–29, Hagley, Soda House, acc. 2198.

103. Brumbley, *Where the Pigeons Slept*, 176–87.

104. The story of the Beneficent Building Association is told in Andrew Heath, *"The Manifest Destiny of Philadelphia": Imperialism, Republicanism, and the Remaking of a City and Its People, 1837–1877* (dissertation, University of Pennsylvania, 2008).

105. Ibid., 198.

106. Beneficent Building Association, *Second Annual Report* (1871), 3, quoted in Heath, *Manifest Destiny of Philadelphia*, 198.

107. Thomas Scharf and Thompson Westcott, *History of Philadelphia, 1609–1884* (Philadelphia: J. H. Everts, 1884), 1458.

108. Boyer, *Urban Masses and Moral Order*.

109. "Through William Penn's 'Low Counties,'" 256.

110. W. Bernard Carlson, *Innovation as a Social Process: Elihu Thomson and the Rise of General Electric, 1870–1900* (New York: Cambridge University Press, 1991).

111. Franklin Institute, Board of Managers Minutes, March 8, 1882.

112. Ibid., January 9, 1884.

113. Ibid., October 8, 1884; "Franklin Institute Exhibition" (1884), Sellers Library, Upper Darby, Scrapbooks, vol. 8, p. 34.

114. Coleman Sellers, "Chairman's Annual Discourse for 1899, on the Transmission of Energy by Electricity," *Proceedings of the American Philosophical Society* 38, no. 159 (read February 3, 1899): 49–71 (quote from p. 57). The minutiae of the Niagara project are documented in Coleman Sellers, Niagara Falls material (1889–99), APS, Sellers Family Papers. For historical accounts of the Niagara Falls power project and its place in the history of electrical system building, see Thomas Hughes, *Networks of Power: Electrification in Western Society, 1880–1930* (Baltimore: Johns Hopkins University Press, 1983); Louis C. Hunter and Lynwood Bryant, *A History of Industrial Power in the United States, 1780–1930*, vol. 3, *The Transmission of Power* (Cambridge, MA: MIT Press, 1991), 260–62.

115. See, for example, Coleman Sellers to I. P. Morris Co., c. 1896; Coleman Sellers to Horace W. Sellers, June 12 and 30, September 19, and October 12, 1897; Coleman Sellers to Edward D. Adams, December 28, 1889; Thomas Edison to Coleman Sellers, January 9, 1890; Elihu Thomson to Coleman Sellers, January 11 and 12, 1895; Nikola Tesla to Coleman Sellers, March 30, 1895; George Forbes to Coleman Sellers, May 24, 1893; Coleman Sellers to Horace Wells Sellers, October 3, 1900, all APS, Sellers Family Papers, Niagara Falls Material.

116. Coleman Sellers to Edward D. Adams, December 28, 1889, Niagara Falls material.

117. Coleman Sellers, "Report to the Board of Directors of the Cataract Construction Company," March 6, 1891, 5, APS, Sellers Family Papers, Niagara Falls material.

118. Sellers, "Chairman's Annual Discourse for 1899," 63.

119. Ibid., 69.

120. Ibid., 63–64.

121. William Unwin to Coleman Sellers, December 13, 1892, APS, Sellers Family Papers; Coleman Sellers to Elihu Thomson, October 20, 1892, APS, Elihu Thomson Papers.

122. W. Newbold to Coleman Sellers, April 9, 1892; Coleman Sellers to W. Newbold, April 9, 1892, APS, Sellers Family Papers.

7. Roots of Decline

1. Agreement between Edge Moor Iron Co. and American Bridge Co., May 12, 1900, Hagley, Soda House, Edge Moor Iron Co. material.

2. Joseph M. Brumbley Sr., *Where the Pigeons Slept: A History* (Wilmington, DE, 1989), 140–41.

3. Philip Scranton, "Large Firms and Industrial Restructuring: The Philadelphia Region 1900–1980," *Pennsylvania Magazine of History and Biography* 116 (1992).

4. Studies of capital mobility and deindustrialization include Barry Bluestone and Bennett Harrison, *The Deindustrialization of America* (New York: Basic Books, 1982); Lloyd Rodwin and Hidehiko Sazanami, eds., *Deindustrialization and Regional Economic Transformation: The Experience of the United States* (Winchester, MA: Unwin Hyman, 1989); Jefferson Cowie, *Capital Moves: RCA's Seventy-Year Quest for Cheap Labor* (Ithaca, NY: Cornell University Press, 1999); Kim Moody, *Workers in a Lean World: Unions in the International Economy* (New York: Verso, 1997); Jefferson Cowie and Joseph Heathcott, eds., *Beyond the Ruins: The Meanings of Deindustrialization* (Ithaca, NY: Cornell University Press, 2003). Studies that push the chronology before World War II and elaborate on other dimensions of class and race include Thomas J. Sugrue, *The Origins of the Urban Crisis: Race and Inequality in Postwar Detroit* (Princeton, NJ: Princeton University Press, 1996); John Cumbler, *A Social History of Economic Decline: Business, Politics, and Work in Trenton* (New Brunswick, NJ: Rutgers University Press, 1989); Philip Scranton, *Figured Tapestry: Production, Markets and Power in Philadelphia Textiles, 1885–1941* (New York: Cambridge University Press, 2002); Scranton, "Large Firms and Industrial Restructuring"; "Deindustrialization: A Panel Discussion," *Pennsylvania History* 58 (1991): 181–211.

5. I have elaborated this argument in greater detail in Domenic Vitiello, "Machine Building and City Building: The Planning and Politics of Urban and Industrial Restructuring in Philadelphia, 1891–1928," *Journal of Urban History* 34, no. 3 (March 2008), from which portions of this chapter are drawn.

6. *Baird et al. v. Rice et al.*, 53 Pennsylvania State Reports (1871), 500; *Baird et al. v. Rice et al.*, 8 Philadelphia Reports (1870–71), 61–67.

7. Lincoln Steffens, "Philadelphia: Corrupt and Contented," in *The Shame of the Cities* (1904; repr., New York: Hill & Wang, 1957), 145.

8. *Ordinances Pertaining to the Removal of Grade Crossings* (Philadelphia, 1902), 18.

9. A. J. Holman, "The Reading Railroad Company vs. the City of Philadelphia" (February 12, 1890), in Citizens' Committee on the Pennsylvania Ave. Track Question, "A Few of the Editorials that Have Appeared from Time to Time in the Daily Papers on the Plan for Depressing the Tracks on Pennsylvania Avenue" (Philadelphia: Citizens' Committee, February 1894).

10. William Sellers to Henry C. Lea, January 16, 1889, Henry C. Lea Papers, Annenberg Rare Book and Manuscript Library, University of Pennsylvania.

11. Mayor Stuart's letter was reprinted in "A Great Work," *Philadelphia Public Ledger*, January 5, 1894, 1.

12. "A Great Municipal Improvement," *Philadelphia Times* (January 15, 1894), in Citizens' Committee, "A Few of the Editorials. . . ," 25–26.

13. "The Reading, the City and Pennsylvania Avenue," *Philadelphia Evening Bulletin*, January 4, 1894, in Citizens' Committee, "A Few of the Editorials. . . ," 11–12.

14. "A Great Work," 5.

15. "Mr. Sellers and the Pennsylvania Avenue Tracks," *Philadelphia Evening Bulletin*, January 15, 1894, in "Few of the Editorials," 27–28.

16. "Pennsylvania Avenue," *Philadelphia Public Ledger*, January 17, 1894, 2.

17. "Objections to the Subway Plan," *Philadelphia Times*, January 17, 1894, in Citizens' Committee, "A Few of the Editorials. . . ," 28.

18. For evidence of the Sellerses' split with the Pennsylvania Railroad interests, see correspondence between William Sellers and Amélie Sellers, summer 1883, Hagley, Soda House, William Sellers & Co. materials.

19. William Sellers to Amélie Sellers, August 7, 1883; Amélie Sellers to William Sellers, August 7, 1883, both Hagley, Soda House, William Sellers & Co. materials.

20. Samuel M. Vauclain, *Steaming Up!* (New York: Brewer & Warren, 1930), 110, 228.

21. The story of the parkway has mainly been told by art and architectural historians, including David B. Brownlee, *Building the City Beautiful: The Benjamin Franklin Parkway and the Philadelphia Museum of Art* (Philadelphia: Philadelphia Museum of Art, 1989);

Cyrus Dezfuli-Arjomandi, *The Benjamin Franklin Parkway in Philadelphia: A Study of the Forces behind Its Creation and Evolution* (dissertation, University of Pennsylvania, 1985); Kenneth Finkel, *Public Architecture and the Emergence of Public Avenues in Philadelphia, 1800–1920* (MA thesis, Temple University, 1978). Also, Albert Kelsey, ed., *The Proposed Parkway for Philadelphia* (Philadelphia, 1902); Andrew Wright Crawford et al., "An Outline of the History of the Fairmount Parkway," Fairmount Park Art Association (Philadelphia, 1922).

22. Thomas L. Hicks, "The Boulevard Well Considered," *Philadelphia Inquirer*, July 2, 1894.

23. Reprinted in Kelsey, *Proposed Parkway for Philadelphia*, 64–73.

24. *Fourth Annual Report of the Commissioners of Fairmount Park*, 51, CPA.

25. Cited in Werner Hegemann and Elbert Peets, *The American Vitruvius: An Architect's Handbook of Civic Art* (1922; repr. New York, 1988), 261; *City and State*, June 19, 1902, and "The Boulevard Proposition Is Gaining Strength . . . ," reprinted in Kelsey, *Proposed Parkway for Philadelphia*, 56, 61–62.

26. George E. Thomas, *The Parkway Chronology* (Philadelphia: GET Associates, 1999).

27. "Park Board Gets Public Squares," news clipping, July 7, 1916, Fairmount Park Art Association Papers, box 55, Benjamin Franklin Parkway Correspondence 1909–13/ 1915–21 folder, HSP.

28. "Joint Bodies Will Plan New Parkway," *Philadelphia Public Ledger*, November 9, 1916.

29. Philadelphia Art Commission, *Annual Report of the Art Jury* (Philadelphia: Art Commission, 1918), 13, University of Pennsylvania Library.

30. "The Pratt & Whitney Company," Pratt & Whitney Papers, Connecticut State Library; *Eleventh Annual Report of the Factory Inspector of the Commonwealth of Pennsylvania for the Year 1900* (Harrisburg, PA: William Stanley Ray, 1901), 114–39, University of Pennsylvania Library; Joseph Wickham Roe, *English and American Tool Builders* (New Haven, CT: Yale University Press, 1916), 178, 222, 255, 259, 273. Niles-Bement-Pond Co., *Machine Tools* (New York: N-B-P, 1903).

31. "Industrial Zoning," *Builders' Guide* 33, no. 51 (December 18, 1918): 689.

32. "The Mayor Monkeying with the Parkway," *Philadelphia Inquirer*, April 4, 1905.

33. *Boyd's Philadelphia Directory for 1925* (Philadelphia: R. L. Polk, 1925), 1162.

34. Albert M. Greenfield Papers, box 27, folder 18, HSP.

35. "Re: Philadelphia Properties of Baldwin Locomotive Works, Revised Schedule of Prices July 1st, 1932" (July 2, 1932); Albert M. Greenfield to Samuel M. Vauclain, July 5, 1932; Samuel M. Vauclain to Albert M. Greenfield, July 6, 1932; "$4,000,000 Arena Projected for Site of Baldwin Works," *Philadelphia Record*, February 1939; "New Rink Rumor Denied by Tyrrell," *Philadelphia Inquirer*, February 9, 1939, all in Albert M. Greenfield Papers, box 92, folder 11, HSP; "School Stadium May Be Built on Baldwin Site," *Philadelphia Record*, 1940, in Albert M. Greenfield Papers, box 100, folder 12, HSP.

36. "The History of Exide Technologies," http://www.exide.com/Media/files/The%20 History%20of%20Exide%20Technologies.pdf (visited March 22, 2012).

37. Joel Schwartz made this argument about postwar redevelopment in Schwartz, *The New York Approach: Robert Moses, Urban Liberals, and Redevelopment of the Inner City* (Columbus: Ohio State University Press, 1993).

38. Philip Scranton, *Endless Novelty: Specialty Production and American Industrialization, 1865– 1925* (Princeton, NJ: Princeton University Press, 1997), 288–89.

39. Answer of Richard S. Dewees, president of the Millbourne Mills Co., to Creditors Petition, Case #2837 (June 18, 1907), National Archives—Philadelphia, RG 21, Eastern Pennsylvania District Court; see also Ellen Cronin, "The Sellers Family of Upper Darby: An Eighteenth and Nineteenth Century American Dynasty of Inventors, Millers and Mechanic-Manufacturers" (typescript in the possession of the author), 23–25.

40. "Memorandum Regarding Edge Moor Iron Company," January 9, 1934, 1, Hagley, Soda House, Wilmington Trust Co., Investment Analysis Files.

41. Ibid., 2.

42. "Earning statements for the ten-year period ending December 31, 1933," Hagley, Soda House, Wilmington Trust Co., Investment Analysis Files; Brumbley, *Where the Pigeons Slept*, 128.

43. Edge Moor Iron Co., Memorandum, November 15, 1935, and "Earning Statements for the Ten-Year Period Ending December 31, 1933," both Hagley, Soda House, Wilmington Trust Co., Investment Analysis Files.

44. Ford, Bacon & Davis Inc., Report on Edge Moor Iron, January 25, 1932, quoted in "Appraisal Value of Edge Moor Iron Company Stocks in the William F. Sellers Estate" (1933), Hagley, Soda House, Wilmington Trust Co., Investment Analysis Files.

45. W. E. S. Dyer to the Stockholders of the Edge Moor Iron Co., February 26, 1935, Hagley, Soda House, Wilmington Trust Co., Investment Analysis Files.

46. "Edge Moor Hears News: The Autonomy of the Picturesque Village Seems Destined to Fade, But a New Plant Is Projected" (July 28, 1935); "Edge Moor Iron Co. Votes for Liquidation" (November 16, 1935); "Famed Iron Works at Edge Moor to Cease Operation: Old Edge Moor Iron, Which Moulded Plate for Brooklyn Bridge, Votes to Liquidate Assets and Dissolve Firm" (November 1935), all unidentified newspaper clippings in HSD, Jeanette Eckman Papers, box 50; *Wilmington Morning News*, July 23, 1935; "Wilmington Will Lose a Notable Industry When Edge Moor Iron Company Liquidates," November 21, 1935, unidentified newspaper clipping in Hagley, Soda House, Wilmington Trust Co., Investment Analysis Files. See also Hagley, Soda House, Edge Moor Iron Co. material.

47. "New Paint Plant Work Is Started: Excavation Begun on DuPont Project at Edgemoor" (c. 1935), unidentified newspaper clipping in HSD, Jeanette Eckman Papers. For more on the liquidation of Edge Moor and related properties, see J. Sellers Bancroft, memorandum, May 22, 1941; final distribution to stockholders, November 12, 1946; "State of Delaware, Office of Secretary of State, Certificate of Dissolution," November 30, 1944, unidentified newspaper clipping, all in Hagley, Soda House, Wilmington Trust Co., Investment Analysis Files; see also Court of Chancery: Helen S. Garrett . . . vs. Edge Moor Iron Co., Hagley, Soda House.

48. Brumbley, *Where the Pigeons Slept*, 155–56.

49. "Coleman Sellers Dies in Bryn Mawr," *Philadelphia Public Ledger*, August 16, 1922.

50. "William Sellers, '23," *Princeton Alumni Weekly*, September 13, 1995.

51. John K. Brown, "When Machines Became Gray and Drawings Black and White: William Sellers and the Rationalization of Mechanical Engineering," *IA: The Journal of the Society for Industrial Archeology* 25, no. 2 (1999): 47.

52. Deed abstracts, plot 3 N 20—26–29, 49, 38, and plot 3 N 21–57, CPA.

53. Scranton, "Large Firms and Industrial Restructuring," 440.

54. "Appraisal Value of William Sellers & Company Stocks in the William F. Sellers Estate" (1933), 11; see also Henry J. Bailey to John H. Shively, May 15, 1934; "Securities Owned by Wm. Sellers & Co. Incorporated, June 30, 1933," "Wm. Sellers & Co., Inc., Condensed Comparative Balance Sheets" (1929–1932), all Hagley, Soda House, Wilmington Trust Co., Investment Analysis Files.

55. "Wm. Sellers & Co. Incorporated, Income Account for Year Ended December 31, 1932," "Wm. Sellers & Co. Incorporated, Income Account for Year Ended December 31, 1935," "Report of the President and Board of Directors for the Year 1935," "Wm. Sellers & Co. Incorporated, Income Account for Year Ended December 31, 1936," "Report of the President and Board of Directors for the Year 1937," "Report of the President and Board of Directors for the Year 1938," "Report of the President and Board of Directors for the Year 1939," "Report of the President and Board of Directors for the Year 1940," all Hagley, Soda House, Wilmington Trust Co., Investment Analysis Files.

56. Albert J. Churella, *From Steam to Diesel: Managerial Customs and Organizational Capabilities in the Twentieth-Century American Locomotive Industry* (Princeton, NJ: Princeton University Press, 1998), 65.

57. Thomas R. Heinrich, *Ships for the Seven Seas: Philadelphia Shipbuilding in the Age of Industrial Capitalism* (Baltimore: Johns Hopkins University Press, 1997), 154.

58. Baldwin Locomotive Works, *The Story of Baldwin* (Philadelphia: F.D. Jacobs, 1943); *Baldwin Locomotive Works Annual Report, 1931*, Hagley Library.

59. Philip Scranton and Walter Licht, *Work Sights: Industrial Philadelphia, 1890–1950* (Philadelphia: Temple University Press, 1986), 194.

60. Heinrich, *Ships for the Seven Seas*, 209; *Machinists Monthly Journal* (September 1921): 771.

61. Howell John Harris, *Bloodless Victories: The Rise and Fall of the Open Shop in the Philadelphia Metal Trades, 1890–1940* (New York: Cambridge University Press, 2000), 275.

62. John K. Brown, correspondence with the author, May 15, 2004.

63. Scranton, *Figured Tapestry*; Walter Licht, *Getting Work: Philadelphia, 1840–1950* (Cambridge, MA: Harvard University Press, 1992), 269 n. 30.

64. On the financial dimensions of the roots of Philadelphia's decline in the 1890s and early twentieth century, see Domenic Vitiello with George E. Thomas, *The Philadelphia Stock Exchange and the City It Made* (Philadelphia: University of Pennsylvania Press, 2010).

65. Ann Markusen, Peter Hall, Scott Campbell, and Sabrina Deitrick, *The Rise of the Gunbelt: The Military Remapping of America* (New York: Oxford University Press, 1991).

66. Harry C. Silcox, *A Place to Live and Work: The Henry Disston Saw Works and the Tacony Community of Philadelphia* (University Park, PA: Pennsylvania State University Press, 1994); William Boekel & Co., "A Century of Service" (text submitted to *Surgical Business*, journal of the Scientific Apparatus Makers Association, April 18, 1968), Collection of Virginia Lehr Budny; Virginia Lehr Budny, Text of video on the Boekel family and Wm. Boekel & Co., c. 1998 (author's collection).

67. Philadelphiana Collection, Fairmount Park—Auto Races 1908–11 folder, Free Library of Philadelphia Prints and Pictures Department.

68. "Plant Is Sold for $1,000,000, Sellers Firm Disposes of Its Property Here," *Philadelphia Evening Bulletin*, May 22, 1947, 37. See also J. Sellers Bancroft et al. to the Stockholders of William Sellers & Co. Inc., December 17, 1942, January 4 and 27, 1943, Hagley, Soda House, Wilmington Trust Co., Investment Analysis Files.

69. Franklin Institute, Board of Managers Minutes, January 12, 1887.

70. Franklin Institute, Annual Report for 1887.

71. Franklin Institute, Annual Report for 1891.

72. Franklin Institute, Annual Reports for 1892–96.

73. Franklin Institute, Annual Report for 1893.

74. "Deaths of the Day: William Sellers Dies after Brief Illness" (January 25, 1905), unidentified newspaper clipping, Union League Archives, William Sellers folder.

75. Clarence S. Bement to Coleman Sellers, December 26, 1892, and William Sellers to Coleman Sellers, December 29, 1892, APS, Sellers Family Papers.

76. Franklin Institute, "Proceedings of the Annual Meeting," *JFI* 137 (February 1894): 160.

77. Franklin Institute, Board of Managers Minutes, August 10, 1898.

78. Franklin Institute, Annual Report for 1899.

79. Maxwell Whiteman, "Philadelphia's Jewish Neighborhoods," in *The Peoples of Philadelphia: A History of Ethnic Groups and Lower-Class Life, 1790–1940*, ed. Allen Davis and Mark Haller (Philadelphia: Temple University Press, 1973), 245.

80. Domenic Vitiello, National Register Nomination, Social Service Building (available at www.arch.pa.state.us).

81. George E. Thomas and David B. Brownlee, *Building America's First University: An Historical and Architectural Guide to the University of Pennsylvania* (Philadelphia: University of Pennsylvania Press, 2000).

82. Margaret Pugh O'Mara, *Cities of Knowledge: Cold War Science and the Search for the Next Silicon Valley* (Princeton, NJ: Princeton University Press, 2004), chap. 4.

83. See Digby Baltzell, *Philadelphia Gentlemen: The Making of a National Upper Class* (New York: Free Press, 1958).

84. For analysis of this stylistic shift in American architecture during this era, see George E. Thomas, *William L. Price: From Arts and Crafts to Modern Design* (New York: Princeton Architectural Press, 2000).

85. Obituaries of Horace Wells Sellers appeared in the *New York Times*, November 27, 1933, and *Philadelphia Public Ledger*, November 28, 1933.

86. For example, Horace Wells Sellers, "Extracts from the Diary of Nathan Sellers, 1776–1778," *PMHB* 26 (1892):191–96; idem, "Letters of Thomas Jefferson to Charles Willson Peale," *PMHB* 28 (1904): 136–54, 295–319, 403–19; idem, "Charles Willson Peale, Artist-Soldier," *PMHB* 38 (1914): 257–86.

87. "William Sellers, '23."

88. "Mrs. Annah Colket McKaig Sellers, 83," *Augusta (GA) Chronicle*, December 14, 2000.

89. Baltzell, *Philadelphia Gentlemen*.

90. Cumbler, *A Social History of Economic Decline*.

91. Sellers, *American Supremacy in Applied Mechanics*, 3.

92. Scranton, "Large Firms and Industrial Restructuring," 464.

93. On the postwar rise of workforce, jobs, and industrial redevelopment strategies, see Guian McKee, *The Problem of Jobs: Liberalism, Race, and Deindustrialization in Philadelphia* (Chicago: University of Chicago Press, 2008); and for university research and related redevelopment, see O'Mara, *Cities of Knowledge*.

INDEX

Note: Page numbers in *italics* indicate illustrations.

abolitionists, 12, 26, 44, 65; Republican Party and, 112; Underground Railroad and, 66–67. *See also* slavery
Adams, Edward, 188
Adams, John, 11
Adams, John Quincy, 51, 52
aesthetics, of machinery design, 158–59
African Americans, 185, 203; Du Bois's studies on, 134, 243 n112; education of, 24, 26, 67; slavery and, 3, 33, 61, 112, 115
Alder, Ken, 222 n21
Allaire Iron Works, 46
Allen, Horatio, 63
All-Night Poker Players club, 119
aluminum industry, 189
Amateur Photographic Exchange Club, 94
American Academy of Political and Social Science, 121–22, 240 n56
American Association for the Advancement of Science, 100
American Bridge Co., 192, 204
American Foundrymen's Association, 132
American Industrial League, 122, 148–49
American Institute, 51, 52, 63, 103, 125
American Institute of Civics, 120
American Iron & Steel Association, 101, 122, 123; merger of, 211; tariff protectionism and, 148–49; Wharton Business School and, 133

American Philosophical Society, 18, 24, 42
American Society of Mechanical Engineers, 93, 105, 110, 129–30
anthracite coal mines, 33, 34, 40, 70
apprenticeship programs, 51, 54–55, 126–29. *See also* education
Apprentice's Library of Philadelphia, 41–42
Arthur, Chester A., 107
Austin, William L., *198*

Baird, Matthew, 83; Centennial Exhibition and, 123; during Civil War, 113, 114
Baldwin, Matthias, 33, 89, 112
Baldwin Locomotive Works, 68, 83, 85, 176; apprenticeships at, 128; Centennial Exhibition and, 123; during Civil War, 114, 115; closing of Philadelphia plant of, 195–99, 202–3, 207–8; Franklin Institute and, 101; labor relations at, 92; photograph of, *201*
Baltzell, Digby, 212, 213
Bancroft, Edward, 50–55, 81; death of, 76; inventions of, 78; wife of, 53, 162. *See also* Bancroft & Sellers Co.
Bancroft, John (Edward's father), 50–51
Bancroft, John Sellers (Edward's son), 93, 116, 177
Bancroft, Joseph (Edward's brother), 50–55, 81
Bancroft, J. Sellers, Jr., 207

Bancroft, Samuel (John's son), 50, 51, 81
Bancroft, William P., 183
Bancroft & Sellers Co., 76, 78–85, 88, 90
Barth, Carl, 127
Beck, James, 199
Beckert, Sven, 118, 240 n42
Beckett, Henry, 85
Bellefontaine & Indiana Railroad, 71–74, 73
Bement, William, 106, 107, 120, 177
Bement & Dougherty Co., 85, 89–90, 176;
    apprenticeships at, 128; during Civil War,
    114, 115; closing of, 195, 196, 200;
    Franklin Institute and, 101; standardized
    bolts of, 103
Beneficent Building Association, 185–87, 211
Bergdoll, Louis, 209
Bergner, Theodore, 93
Bernstein, Iver, 240 n42
Bessemer steelmaking process, 138, 139, 141
Bethlehem Steel Co., 137, 149, 208, 247 n89
Biddle, Nicholas, 55, 56
Biggs, Lindy, 234 n56
Blankenburg, Rudolph, 199, 200
Blodgett, Lorin, 121
Bringhurst, Edith Ferris, 183
Brinley, Charles, 141–47, 149, 151
Brooklyn Bridge, 146, 165
Brooks, James C., 116
Brown, John K., 91, 92, 113, 139, 244 n6
Brown & Sharpe Co., 78; industrial
    standardization by, 105–6; military contracts
    of, 153
Brunel, Marc Isambard, 49
Burnham, Daniel, 195
Bush Hill district, 84–88, 99, 118, 176–77;
    deindustrialization of, 194–96, 199, 207;
    illustrations of, 86, 87, 201; vocational high
    school in, 129
Butcher, William, 138–41

Calhoun, John, 68
Cambria Steel Co., 148
Cammell & Co., 151
canals, 9, 20, 23–24, 28–29, 51, 68. See also
    infrastructure
Carbon Run Improvement Co., 81
Carborundum Co., 189
Cardington factory, 36–37; closing of, 55;
    locomotives at, 39, 53–54
Carey, Henry, 107, 112, 122
Carnegie, Andrew, 141, 149, 151; Keystone
    Bridge Co. of, 140–41, 160, 192; military
    contracts of, 154
Cataract Construction Co., 188

Centennial Exhibition (1876), 110, 123–24,
    133; arms manufacturers at, 136; building
    designs at, 159; Fairmount Park and, 168,
    171. See also exhibitions
Chandler, Alfred D., 244 n6
Chase, Salmon, 122
chemical companies, 24, 116, 122, 167, 181
Chicago, 71; Columbian Exposition in, 163,
    199; Mercantile Exchange of, 118
Chicago School architects, 175
Children's Aid Society, 211
Chilean railroad, 83
cholera, 25–26
Chrome Steel Co., 141
Churella, Albert, 207–8
Cincinnati, 56–61, 71, 72
Cincinnati Southern Railroad, 163
cinema, development of, 94
Citizens' Committee of '95 for Good City
    Government, 120
Citizens' Municipal Reform Association, 120
City Hall of Philadelphia, 175
civil service, professionalization of, 120, 165
Clark, Clarence (Edward's son), 107, 119, 120;
    Butcher Steel Works and, 140; Midvale Steel
    Co. and, 137, 142, 151; residence of, 172;
    Frederick Taylor and, 146; Western Home
    for Poor Children and, 186
Clark, Edward, 140–42, 154
Clark, William, 24
coal mining. See anthracite coal mines
Coal Run Improvement Railroad, 70
Coleman, Joseph, 18, 27
Colt, Samuel, 153
Columbian Exposition (1893), 163, 199
computer industry, 212
Consolidated Machine Tool Co., 209
Constitutional Convention, 23
Cooke, Jay, 111; bankruptcy of, 142;
    Centennial Exhibition and, 123
Copeland, Robert Morris, 178, 181, 183
Corliss, George, 55
Coxe, Tench, 23
Cramp, Charles, 100, 115
Cramp shipyard, 137; apprenticeships at, 128;
    closing of, 208; military contracts at, 149–50;
    Taylorism at, 247 n89; workforce at, 154
Crédit Mobilier, 2, 68
Cresson, John, 130
Cumbler, John, 213

Daniels, Thomas, 249 n38
Davenport, Russell Wheeler, 143–44, 146, 149,
    151

Dawson, Andrew, 128, 245 n50
deindustrialization, 194–204, 221 n20. *See also* industrialization
depression: of 1830, 99; of 1930s, 193, 208
Dewees, Richard, 205
Disston, Henry, 82, 182
Disston Saw Works, 150, 181, 209
Dobson textile mills, 181
Donkin, Bryan, 49
Dougherty, James, 100
Drexel, Anthony, 107, 172; Centennial Exhibition and, 123; war bonds of, 111; Western Home for Poor Children and, 186
Du Bois, W.E.B., 134, 243 n112
R.G. Dun credit agency, 90, 116; on Butcher Steel Works, 140, 141; on Edge Moor Bridge Works, 163, 164; on Midvale Steel Co., 142
Du Pont family, 52, 183; gunpowder mills of, 150, 162; paint factory of, 206
Dyer, W.E.S., 206
dynamometer, 81

Eads, James, 140–41
Eagle Screw Co., 102–3
Eastern State Penitentiary, 37, 87
East Tennessee & Georgia Railroad, 68, 69
economic development, 22–23, 77–78; geographic factors in, 2, 9–10, 106, 161–62; networks of, 4–8, 107–12, 117–25, 162, 175–87, 191
Edge Moor Bridge Works, 156, 159, 163–64, 175, 192
Edge Moor Iron Co., 110, 124, 126, 127, 137; closing of, 204–6; geographic advantages of, 161–62; operations at, 147–48, 159–61; University of Pennsylvania and, 131
Edge Moor village, 157, 178, 181–85, *184*
Edison, Thomas, 188
education, 185; of African Americans, 24, 26, 67; apprenticeship programs and, 51, 54–55, 126–29; of engineers, 40–48, 42, 54–55, 57, 110; libraries and, 41–42, 121; of pharmacists, 42; private, 44–45; public, 41–45, 47, 62, 67
educational-industrial complex, 125–35. *See also* military-industrial complex
electricity, 24–25, *109*, 133, 156–57, 187–90, 253 n114
Electric Storage Battery, 203
elevators, 79, 175–76
Elkins, William, 119, 171, 200
Ely, Theodore, 95

Emancipation Proclamation, 115
Engineers' Club of Philadelphia, 123, 129
environmentalism, 183, 186; city parks and, 166–75, 183–85, *184*, 190–91; public health and, 25–26, 163, 167, 171–75, 249 n38. *See also* water pollution
Erie Canal, 28–29, 51. *See also* canals
Evans, Oliver, 35–37, 43
exhibitions, 63; of electrical technology, 187; at Franklin Institute, 81, 99–100, 123–25, 133, 158; world fairs and, 125, 163, 199. *See also* Centennial Exhibition

Fairbanks, Bancroft & Co., 55
Fairmount Park, 166–69, *170*, 186; Commission of, 120, 124, 178, 187
Fairmount Park Art Association, 169
Fairmount Parkway, 199–201, *201*, 211, 214
Federalists, 223 n47
Fels, Samuel, 211
Felton, Samuel, 138
Ferris, Mary, 77, 81, 116, 162
fire engines, 29, 32–35, *36*, 42; hoses for, 33, 81–82
Fraley, Frederick, 107, 130, 133
Franklin, Benjamin, 187; American Philosophical Society and, 18; *Pennsylvania Gazette* of, 16; statue of, 211
Franklin Institute, 42–43, 46, 122–23, 194; decline of, 209–11; electrical research at, 133, 187, 188; exhibitions at, 81, 99–100, 123–25, 133, 158; federal lobbying by, 154; financial problems of, 99–100; industrial standardization and, 77, 103; on metric system, 105; photography course at, 95; sanitary engineering by, 174; technical training at, 127–28, 130
Franklin Sugar Refineries, 151
Freedley, Edwin: *History of American Manufactures* by, 96, 99; *Leading Pursuits and Leading Men* by, 93; *Philadelphia and Its Manufactures* by, 89–90
Free Produce Association of Friends, 67
Free-Soil Party, 66
fugitive slave law, 113
Furness, Horace, 130

Gallman, J. Matthew, 238
Galveston Rope & Twine Co., 164
Garrett, Eli, 162, 163
Garrett family, 206
German immigrants, 114, 115. *See also* immigrant workers
Gibens, Anna, 14

Giffard Injector, 95–96
Giguere, George, *3*
Gilded Age, The (Twain & Warner), 1–2, 65, 69, 74–75, 165, 217n3
Gilpin, Joshua, 49
Girard, Stephen, 30–31
Girard College, 37
GlaxoSmithKline Corporation, 203
global economy, 125, 128–29, 240n41, 243n4; Centennial Exhibition and, 136; infrastructure and, 156, 163; protectionist tariffs and, 37, 51, 112, 122, 133, 135, 148–49
Globe Rolling Mill and Wire Works, 57–58, 63–64
Gold Rush (1849), 64
gold standard, 123
Grant, Ulysses S., 114, 122
Great Awakening, Second, 53
Great Depression, 193, 208
Green, Duff, 2, 67–70
greenback monetary policy, 123
Greenwood, Miles, 63
Greenwood, Ronald, 142
Gun Foundry Board, 153
gunpowder manufacturing, 11, 16, 20, 150
Guyana, 163, *164*

Haasz, Amélie, 116–17, 183, 198
Halsey, Frederick, 158
Harrah family, 154–55
Haupt, Herman, 83
Hazard, Erskine, 33, 34
Heath, Andrew, 89, 186
Heinrich, Thomas, 150
Hewitt, Abram, 108
Hill, George, 38
Hill, Hannah, 59
Hill, Peter, 38
Holley, Alexander, 105, 139, 148, 163
Holmes, Oliver Wendell, 94
Hoopes, Barton, 54, 85
Hoopes & Townsend Co., 85, 207
Hounshell, David, 236n98
Hunter, Louis C., 232n15
Huston, Samuel, 140, 142, 154
Hydraulion (fire engine), 35, *36*, 49

Ibotson, Richard, 49
immigrant workers, 47, 50–51, 53; German, 61, 62, 114, 115; Irish, 61, 62; Know-Nothing Party and, 62; settlement houses for, 122
industrialization, 2–6, 29–40, 77–78, 89–90; in antebellum South, 65–70; ecology of, 78–79, 221n19; education for, 40–45; environmental

concerns with, 165–66; innovation and, 90–100, 106. *See also* deindustrialization; economic development
industrial revolution, second, 6, 138, 147–49, 155–57, 187–90, 204, 215
infrastructure, 20, 75, 156–60, 163; canals, 9, 20, 23–24, 28–29, 51, 68; energy, 34, 40, 187–90; for interregional trade, 47, 56, 62–63, 72; investments in, 4–7, 10, 25, 68, 175–76, 213; social reforms and, 12, 18, 23, 111, 190–91; turnpikes, 24; urban development and, 20, 58, 135, 162, 165
innovation, 90–100, 106, 122–23
institutional complexes, 8–9, 99–106, 221n18; educational, 125–35. *See also* military-industrial complex
interchangeable parts, 9, 77, 91–93, 98–102, 236n98; for building design, 159–60. *See also* standardization
International Association for Testing Materials, 131

Jackson, Andrew, 23, 37, 68
Jacobs, Jane, 9
James, C.F., 81
James, Edmund, 121, 122, 165
Jefferson, Thomas, 23
Jenney, William, 175
Johnson, John, 200
Jones, Thomas, 33
Justice, Philip Syng, 139–40

Katz, Michael, 41
Keating, William H., 42
Kelley, William "Pig Iron," 77, 85, 122; on Centennial Exhibition, 123; on Pennsylvania Railroad's abattoir, 174
Kelvin, Lord, 189
Kensington Iron & Steel Works, 83
Kentucky River Bridge, 163
Kershaw Steam Mill, 82
Keystone Bridge Co., 140–41, 160, 175, 192
kinematoscope, *94*
Kneass, Strickland (father), 83, 124–25, 132–33
Kneass, Strickland L. (son), 96, 126
Know-Nothing Party, 62
Krupp steelworks, 139, 152
Kuhn, Hartman, 85, 88

Latrobe, Benjamin, 26
Lawrence, Josiah, 57, 64
Lea, Henry Charles, 107, 119–22, 196
Lehigh Navigation Co., 34

Lewis, Enoch, 83, 126–27
Lewis, Meriwether, 24
Lewis, Robert, 7, 251 n70
Lewis, Wilfred, 93, 126–27, 147
libraries, 41–42, 121
Licht, Walter, 241 n76
Lincoln, Abraham, 114, 115
Lippincott, Charles, 124
Lippincott, J.B., 107; Ridley Park and, 178, *179*; Wharton Business School and, 133
Logue, Ed, 203
Louisiana Purchase Exposition, 125
Lowell, James Russell, 44
Lukens, Isaiah, 33
Lukens, John, 18, 19
Lyon, Pat, 32, 33

malaria, 143, 144, 173
Marburg, Edgar, 131, 165, 175
Mascart, Éleuthère, 189
Master Mechanics Association, 105
Matteawan Manufacturing Co., 51–52, 54
Matthews, John, 49
McCaffery, Peter, 119
McMichael, Morton, 107, 119, 120; Fairmount Park Commission and, 167–69; Philadelphia Industrial League and, 122
McNichol, "Sunny Jim," 119
Mechanics Institute (Cincinnati), 61
Medico-Chirurgical College, 200
Merrick, J. Vaughan, 100, 103, 119, 120; University of Pennsylvania and, 130, 131
Merrick, Samuel Vaughan, 42, 43, 89, 113, 114
Merrick & Sons, 101, 114
Merrick & Towne, 81
Metcalfe, Henry, 153
metric system, 105
Mexican War, 62
Midvale Steel Co., 110, 127, 137–49, 210; drawing of, *143*; foreclosure of, 142; military contracts of, 144–45, 149–55, *152*; Siemens process at, 136; Taylorism at, 146–47; wages at, 145–47; workforce at, 146, 154
Mifflin, Thomas, 23
military-industrial complex, 104, 136–38, 149–55, 190; Civil War and, 101, 102, 111, 115; educational institutes and, 125–35; Revolutionary War and, 20–21. *See also* networks
Millbourne Mills Co., 31, 43, 127, 137, 176; closing of, 204–5
Misa, Thomas, 149, 175
Mogeme, Henri, 46–49, 53
Monitor (ironclad ship), 115

Monroe, James, 68
Moore, James, 54
Morgan, J.P., 192, 208
Morris, Isaac P., 113
Morris, Israel, 35
Morris, Robert, 11, 23
I. P. Morris company, 81, 85, 86, 89; apprenticeships at, 128; during Civil War, 115; Niagara power plant and, 189; during Spanish-American War, 150; Taylorism at, 247 n89
Morrison Steam Hammer, 96
Morton, Henry, 101, 130
Moses, Robert, 203
Muller, Ted, 7

Nashville (TN), 58, 70
Nashville Manufacturing Co., 81
National Academy of Science, 101
networks, 30–31, 38–39, 55, 81; of economic development, 4–8, 107–12, 117–25, 162, 175–87, 191; educational-industrial, 125–35; institutional, 8–9, 99–106, 221 n18; of social reform, 26, 110, 117–22, 125, 185–87, 190–91; of textile manufacturers, 38–39, 50–51. *See also* military-industrial complex
Niagara Falls Power Company, *109,* 156–57, 187, 253 n114
Niles, Jonathan and James, 71
Niles-Bement-Pond Corporation, 200, 211
Niles & Co., 70–73, 82
Norris, William, 63
Norris locomotive works, 83, 85
Norristown Railroad, 84, 140

Ohio Mechanics Institute, 57
Ohio & Mississippi Railroad, 72
Olmsted, Frederick Law, 166, 178
ordnance manufacturing, 115, 136–37, 149–54, *152. See also* military-industrial complex
Oregon territory, 62
Outerbridge, Alexander, Jr., 95, 153
Owen, Robert, 43

Panama Canal, 137
Panama Railroad, 1–2, 63, 65, 70
panic: of 1837, 55; of 1857, 72, 112; of 1873, 142, 159; of 1893, 204
paper manufacturing, 11, 21, 29, 33, 36, 49
parks, city, 166–75, 183–85, *184,* 190–91
Parrish, Rachel, 49
Parry, Charles, 100, 103, 124
Patterson, Robert, 33
Peale, Charles W. (accountant), 101

Peale, Charles Willson (artist), 2, *13*, 24, 43, 188
Peale, Franklin, 33, 82
Peale, Louisa, 64
Peale, Sophonisba. *See* Sellers, Sophonisba
Peale, Titian, 94
Penn, William, 14
Pennock, Abraham, 33–35, *36*, 44, 66–67
Pennsylvania Company for Insurance, 162
Pennsylvania Fiscal Agency, 68
Pennsylvania Industrial League, 120, 122, 123, 133
Pennsylvania Museum and School of Industrial Art, 110, 130
Pennsylvania Railroad, 102, 198; abattoir of, 172–74, *173*; bridges for, 160–61, *161*, 163; during Civil War, 111–12; injectors for, 95–96; Midvale Steel contracts of, 143; standardized screws for, 104–5
Pennsylvania Society for Encouragement of Manufactures and the Useful Arts, 12, 24
Pennsylvania Society for the Improvement of Roads and Inland Navigation, 23
Pennsylvania Society for the Promotion of Public Economy, 40–41
Pennsylvania Society for the Protection of Animals, 134
Pennsylvania Steel Co., 138, 139
Pepper, William, 107
Perkins, Jacob, 32, 33, 42
pharmaceutical companies, 82, 203
Philadelphia, Wilmington & Baltimore Railroad, 84, 102, 112, 138, 183
Philadelphia College of Pharmacy, 42
Philadelphia Community College, 209
Philadelphia Interests (investment group), 138, 140
Philadelphia Manufacturers' Club, 123, 211
Philadelphia Museum of Art, 200
Philadelphia Social History Project, 251 n77
Philadelphia Social Science Association (PSSA), 120–22, 211–12; technical education and, 126; urban planning and, 157, 186; Wharton Business School and, 134
Philadelphia Society for Promoting Agriculture, 12, 24
Philadelphia Textile School, 128
Phoenix Bridge Company, 160, 175
Photographic Society of Philadelphia, 94
photography, *94*, 94–95
Pittsburgh Reduction Co., 189
Pittsburgh Steel Casting Co., 153
Poole, Elizabeth, 66, 81
Poole, John Morton, 51–54

Poole, Sarah, 50, 51, 53, 54
Poole, William, 19, 31, 50
Powers & Weightman company, 181
Pratt & Whitney, 78; industrial standardization by, 105–6; military contracts for, 153; Niagara power plant and, 189
Pred, Allan R., 220 n14
Price, Eli Kirk, 88, 89, 167–71
Progressivism, 110, 117–20; Philadelphia Social Science Association and, 121–22; public health and, 173–75; suburban development and, 181–87, 190–91, 195; urban reform and, 6, 10, 110, 125, 157, 195
protectionism. *See* tariffs
PSSA. *See* Philadelphia Social Science Association
Pullman (IL), 182, 183
Pusey & Jones company, 81

Quakers, 14, 19, 26, 149; "fighting," 20, 22, 149; revivalists and, 53; schools of, 31, 44; slavery and, 3

race riots (1952), 203
railroad(s), 9; Bellefontaine & Indiana, 71–74, *73*; bridge manufacturing for, 160–63, *161*; in Chile, 83; Cincinnati Southern, 163; Coal Run Improvement, 70; East Tennessee & Georgia, 68, 69; industrial standardization of, 102, 104–5; injectors for, 95–96, 124–25, 132–33; Norristown, 84, 140; Ohio & Mississippi, 72; in Panama, 1–2, 63, 65, 70; within Philadelphia, 83–85, *84*; Philadelphia, Wilmington & Baltimore, 84, 102, 112, 138, 183; River Front, 183, 192; transcontinental, 1, 64–65, 69; wheels for, 136, 139–40, 143, 145, 148. *See also* Pennsylvania Railroad
Reading Railroad, 83–84, 196–97; Midvale Steel and, 144; rail contracts of, 140, 141
Reaney & Neafie Co., 83
Reform Club, 120
Rensselaer Institute of Technology, 126
Rensselaer Polytechnic Institute, 163
Republican Club, 112
Republican Party, 117–20, 129; on deindustrialization, 194–200, 203; Union League and, 119, 187
Richards, Thomas, 130
Ridley Park, 178–81, *179*, *180*
Rittenhouse, David, 11, 18, 19, 23
River Front Railroad, 183, 192
Roberts, George, 198
Robson, Charles, 86, 87, 98

Roe, Joseph, 79, 158, 160
Rogers, Fairman, 133
Ronaldson, James, 51
Rosengarten family, 116
Rush, Benjamin, 32
Rush & Muhlenberg Co., 85

Salt, Titus, 182, 183
Saturday Club, 107–10, 113, 118–19, 122;
    Centennial Exhibition and, 123; dissolving
    of, 211; Fairmount Park and, 178; University
    of Pennsylvania and, 130, 133
Scharf, Thomas, 185
Schenck, Peter, 51
Schuyler, David, 249 n37
Schuylkill River Bridge, 163
Scott, Thomas, 107, 111–12, 140–41
Scranton, Philip, 9, 104, 105, 128, 238 n7,
    244 n6
screw threads, standardization of, 101–6
Second Great Awakening, 53
second industrial revolution, 6, 138, 147–49,
    155–57, 187–90, 204, 215
Sellers, Alexander (William's son), 116, 183, 207
Sellers, Amélie Haasz (William's second wife),
    116–17, 183, 198
Sellers, Charles (Coleman's son), 2, 35–36;
    education of, 42, 43, 44; locomotive manu-
    facturing by, 39; in South and West, 55–75;
    Upper Darby Institute and, 43
Sellers, Charles Coleman (Horace's son), 217 n1
Sellers, Coleman (Nathan's son), 6, 10, 12, 33;
    Apprentice's Library and, 41–42; businesses
    of, 29–30, 35–36, 38, 81; death of, 38, 39;
    education of, 31–32; wife of, 24, 32, 43, 44
Sellers, Coleman, II (Coleman I's son), 77–79,
    93; in Cincinnati, 60–62, 82; during
    Civil War, 112–13, 115; education of, 44;
    European trips of, 63, 152, 160, 182, 188;
    Franklin Institute and, 100; legacy of, 204;
    on machines as "slaves," 123; on military-
    industrial complex, 136–37, 190; Niagara
    Falls Power Co. and, 109, 187–90; on Penn's
    veterinary school, 134–35; photographic
    experiments of, 94, 94–95; portrait of,
    109; residences of, 172, 177; retirement of,
    122, 126, 188; salary of, 116; as teacher, 95,
    129–30
Sellers, Coleman, Jr. (Coleman II's son), 204;
    death of, 206; education of, 126; Engineers'
    Club of Philadelphia and, 123; Franklin
    Institute and, 210–11; on military contracts,
    153; photograph of, 94

Sellers, David (John I's son), 11–12; apprentice-
    ship of, 17; bank stocks of, 23–24; death of,
    32; enterprises of, 30–31, 36; wife of, 18
Sellers, David, Jr. (David's son), 31
Sellers, Frances (John III's daughter), 162
Sellers, George (Samuel I's brother), 14
Sellers, George (Nathan's brother), 17
Sellers, George (William I's brother), 162
Sellers, George Escol (Coleman I's son):
    education of, 43, 44; European trips of, 2–3,
    49–50, 138; and The Gilded Age, 1–2, 65,
    69, 74–75, 165; on his grandparents, 27–28;
    on his parents, 33; inventions of, 73–74;
    locomotive manufacturing and, 39, 63,
    65, 95; memoirs of, 2–3, 74; portrait of, 3;
    retirement of, 2, 74–75; in South and West,
    55–75; Upper Darby Institute and, 43; wife
    of, 49, 60
Sellers, Horace (Coleman II's son), 94, 187,
    188, 193, 212–13
Sellers, Howard (John Jr.'s son), 126
Sellers, James (David's son), 36
Sellers, John I (Samuel Jr.'s son), 16–19; as con-
    stable, 18; after Revolutionary War, 23; dur-
    ing Revolutionary War, 20–21; wife of, 17
Sellers, John, II (John I's son), 17, 18, 31
Sellers, John, III (John II's son), 19, 31; aboli-
    tionism, 66
Sellers, John, Jr. (John III's son), 59, 77–79, 93,
    116–17; during Civil War, 112–16; death of,
    204–5; at Edge Moor, 162; environmental
    concerns of, 166, 169, 172–74, 173; at Mill-
    bourne Mills, 137; residences of, 172, 177;
    Western Home for Poor Children
    and, 186
Sellers, Mary (John III's daughter), 53, 162
Sellers, Morris (Charles's son), 62
Sellers, Nathan (John I's son), 3, 9–12; bank
    stocks of, 23–24, 38; as city councilor,
    25–26; death of, 12; enterprises of, 29–31,
    38; as notary, 17; portrait of, 13; retirement
    of, 28, 32; after Revolutionary War, 23;
    during Revolutionary War, 20–22; view
    of progress of, 17–18, 27–28; wife of, 18,
    27–28
Sellers, Richard (William's son), 116
Sellers, Samuel, I (John I's grandfather), 14
Sellers, Samuel, Jr. (Samuel I's son), 16
Sellers, Samuel I (John I's son), 17, 21
Sellers, Samuel II (David's son), 36, 40–42
Sellers, Sophonisba "Sophy" Peale (Coleman's
    wife), 24, 32, 43, 44; Cincinnati trip of,
    59–61; on slave labor, 66, 67

Sellers, William (John III's son), 6, 9, 10, 107–11; apprenticeship of, 54–55; cartoon of, *198*; Centennial Exhibition and, 123–25, 168; during Civil War, 112–16; death of, 195, 210; European trip of, 138–39; Fairmount Park and, 166–72; legacy of, 204–5; Midvale Steel Co. of, 137–49; portrait of, *108*; residences of, 114–15, 172, 177, 182–83, *184*, 195, 199; retirement of, 183; on universal suffrage, 120; University of Pennsylvania and, 110, 130–35; wives of, 77, 116, 162, 177, 182–83, 198

Sellers, William, II (William I's grandson), 206, 213

Sellers, William Ferris (William I's son), 77, 126, 204, 205

William Sellers & Co., 55, 72, 76–79, 90; apprenticeships at, 126–27; Centennial Exhibition and, 123–25; closing of, 206–7, 209; Edge Moor and, 159–60, 162; Franklin Institute and, 101; hydraulic tools at, 139; incorporation of, 122; injectors of, 95–96, 124–25, 132–33; military contracts of, 149–55, *150, 152*; photograph of, *80*; rationalization of work at, 90–99, 156; standardized screw threads of, 101–6; steam hammer of, 96–98, *97*; workforce at, 91–93, 115

Sellers Family Association, 212–13

Sellers Landing (IL), 74

Sellers & Pennock, 33, 35, *36*, 46, 48, 81–82

Sellers & Sons, 36–37, *39*, 43, 49, 52, 55, 81

settlement houses, 122

Sharp, Steward & Co., 95

Shedaker, Hudson, 174

Siemens steelmaking process, 136, 140, 141; tariffs on, 148–49

Silcox, Harry, 182

silk industry, 19–20. *See also* textile manufacturing

Simpson, George, 112–13

skyscrapers, 175–76, 251 n76

Slaton, Amy, 251 n76

slavery, 3, 33, 61, 112, 115. *See also* abolitionists

Smith, Elvino, 233 n44

Smith, Merritt Roe, 236 n98

Smith, William, 18

Smith Kline French (SKF) Corporation, 203

Smithsonian Institution, 100

social reform, 6, 10, 211–12, 247 n2; infrastructure planning and, 12, 18, 23, 111; networks of, 26, 110, 117–22, 125, 185–87, 190–91

Society for the Institution and Support of Sunday Schools, 12, 26

Society for the Prevention of Cruelty to Animals (SPCA), 26, 134, 135

Society of Mysterious Pilgrims, 119

Spanish-American War, 137, 150

Spring Garden district, 87–89, 177, 203

Spring Garden Gas Works, 174

Spring Garden Institute, 128, 129, 131

standardization: of design, 91–93, 131, 156–60; of screw threads, 101–6. *See also* interchangeable parts

Standard Oil Company, 119, 171

steam hammer, 96–97, *97*

Steffens, Lincoln, 196

Stevens Institute of Technology, 129, 130, 132, 146

Stillé, Charles, 107, 121, 123

Stotesbury, E.T., 200

Stuart, Edwin, 197

suburbanization, 16, 181–87, 190–91, 195, 202. *See also* urban planning

Swedenborgianism, 22, 26, 28

tariffs, protectionist, 37, 51, 112, 122, 133, 135, 148–49

Tasker & Morris iron foundry, 81

Taylor, Frederick Winslow, 138, 146–47, 149, 153, 155; at Bethlehem Steel, 247 n89; rationalization of work and, 90–99, 156

temperance movement, 3, 28, 41, 53, 67

Tesla, Nikola, 188

textile manufacturing, 12–16, 29, 128, 181; during Civil War, 111–13; machinery for, 82; networks of, 38–39, 50–51; silk, 19–20; women in, 128

Thompson, J. Edgar, 83, 112, 138, 140–41

Thomson, Elihu, 190

Thorne, William, 93, 127

Towne, Henry (John's son), 127

Towne, John, 103, 112, 114, 127; University of Pennsylvania and, 131, 134

Trotters company, 30–31

Trumbauer, Horace, 200

Tuberculosis Society, 211

turnpikes, 24. *See also* infrastructure

Turrettini, Théodore, 189

Twain, Mark: *The American Claimant* by, 75; *The Gilded Age* by, 1–2, 65, 69, 74–75, 165, 217 n3

Tweedale, Geoffrey, 141, 143

Tyler, John, 68

Underground Railroad, 66–67. *See also* abolitionists

Union Carbide, 189

Union League, 110, 112–15, 117, 122, 211; Centennial Exhibition and, 123; Republicans of, 119, 187

Unisys Corporation, 212

United Gas Improvement Co., 122, 203

universal suffrage, 120

University of Pennsylvania, 18, 24, 110, 194, 212; business school of, 133–34; engineering at, 126; medical school of, 12, 130–31; power plant of, *132*; science department of, 42, 101, 131; social reformers and, 121–22; urban planning at, 165; veterinary school of, 134–35

Unwin, William, 189, 190

Upper Darby, 31–32, 36, 43, 54, 127

urban planning, 25, 110, 156–57, 218 nn5–6; deindustrialization and, 194–204; engineering and, 4, 9, 165; geographic factors in, 2, 9–10, 106, 161–62; market model of, 186; nation building and, 20–29; parks and, 166–75, 183–85, *184*, 190–91. *See also* suburbanization

urban reform. *See* social reform

Ure, Andrew, 79

U.S. Mint, 33, 48, 82, 85, 176, 209; Andrew Jackson and, 37; Revolutionary War and, 21

U.S. Navy, 82, 132; guns for, 136, 151; industrial standardization by, 104; military contracts of, *150*, 150–53

U.S. Postal Service, 33

U.S. Shipbuilding Corporation, 208

U.S. Steel Corporation, 204, 211

Vansant, I.L., 86–87

Vauclain, Samuel, 177, 198–99, 202

veterinary medicine, 134–35

Vice and Immorality Committee, 41

Warner, Charles Dudley, 1–2, 65, 69, 74–75, 165, 217 n3

War of 1812, 30

War of Independence, 20–29, 193

Washington, George, 11, 193

water pollution, 25–26, 163–73, 200, 224 n62, 250 n38. *See also* environmentalism

Wells, Cornelia, 70

Welsh, John, 107, 121, 134; Centennial Exhibition and, 123; Fairmount Park and, 167; University of Pennsylvania and, 130

Western Home for Poor Children, 186

Westinghouse Electric Co., 189

Wharton, Joseph, 112, 133–34, 138; Bethlehem Steel Co. of, 137, 149; military contracts of, 154; Philadelphia Industrial League and, 122

Wharton family, 116

Whetstone, John, 63, 65, 70

White, Josiah, 33, 34

Whitney, Asa, 83, 85, 112; Centennial Exhibition and, 123; during Civil War, 113, 114

Asa Whitney & Sons, 148; Butcher Steel Works and, 140; closing of, 207

Whittier, John Greenleaf, 44

Whitworth, Joseph, 103, 104, 151, 152

Whitworth & Co., 147, 151, 153

Widener, Peter, 119, 200, 211

Wills Asylum (eye hospital), 88

Wilson, Joseph, 163, 175, 196, 210

Wiltbank, John, 35

women's suffrage, 66–67

world's fairs, 123–25, 163, 199. *See also* exhibitions

Wrege, Charles, 142

Wright, James, 154

Yale, Linus, 127

yellow fever, 25

Zeller, Theodore, 104

zoning codes, 200

CPSIA information can be obtained
at www.ICGtesting.com
Printed in the USA
BVOW06*1223071117

499764BV00012B/202/P